SCIENTIFIC AMERICAN EXPLORES BIG IDEAS

Discoveries in Astronomy

Published in 2025 by Scientific American Educational Publishing
in association with **The Rosen Publishing Group**
2544 Clinton Street, Buffalo NY 14224

Contains material from Scientific American®, a division of Springer Nature America, Inc., reprinted by permission, as well as original material from The Rosen Publishing Group®.

First Edition

Scientific American
Lisa Pallatroni: Project Editor

Rosen Publishing
Michael Hessel-Mial: Compiling Editor
Michael Moy: Senior Graphic Designer

Cataloging-in-Publication Data
Names: Scientific American Editors (New York, New York), editor.
Title: Discoveries in astronomy / The Editors of Scientific American.
Other titles: Scientific American explores big ideas.
Description: First edition. | New York : Scientific American Educational Publishing, 2025. | Series: Scientific American explores big ideas | Includes bibliographical references and index. | Audience: Grades 10-12 |
Identifiers: LCCN 2024015181 | ISBN 9781725351813 (library binding) | ISBN 9781725351806 (paperback)| ISBN 9781725351820 (ebook)
Subjects: LCSH: Astronomy.
Classification: LCC QB43.3 .D57 2025 | DDC 520–dc23/eng/20240508
LC record available at https://lccn.loc.gov/2024015181

Manufactured in the United States of America
Websites listed were live at the time of publication.

Cover: muratart/Shutterstock.com

CPSIA Compliance Information: Batch # CSSA25.
For Further Information contact Rosen Publishing at 1-800-237-9932.

CONTENTS

INTRODUCTION

As a science, astronomy explores the furthest objects conceivable to humans: celestial bodies millions or more light-years away, or historic traces of the most distant states of the universe. This research has been aided not only by technological advances, but by ingenious ways of using them. The different wavelengths of the electromagnetic spectrum, of which visible light is just a small piece, have allowed for telescopes to pull great details from tiny patches of the sky—including planets orbiting distant stars. Gravitational wave astronomy, among the newest and most technically challenging, gives us even more ways to observe the universe.

The articles in this title are organized by closeness to us. The first section, "Exploring the Solar System," highlights how the search for life and study of how planets form remain unanswered questions in our backyard. The second section, "Hunting Exoplanets," covers the last decade's highlights of the thousands of new planets discovered. These include planet types far different from our own solar system, raising new questions about what a habitable planet might look like. The third section, "The Milky Way and Beyond," presents new findings on a wide variety of strange celestial objects: black holes, neutron stars, brown dwarfs, fast radio bursts, and more. The fourth section, "Cosmology Explains the Universe," dives into the most remote questions: the universe's origins and structure, the nature of its expansion, and the dilemma presented by still-undetected dark matter and energy. The final section, "Policy Behind Astronomy," brings us back to Earth to consider the people who shape how astronomy research is conducted. Here you'll see the challenges and debates that will shape the practice for decades to come.

As you read, you'll learn more about the strange objects that populate our cosmos, and see the familiar ones in a new light. Hopefully, you'll marvel just as other readers have, not only at the scope of space research, but at the efforts people have made to advance it.

Section 1: Exploring the Solar System

Venus Might Host Life, New Discovery Suggests

By Adam Mann

There is something funky going on in the clouds of Venus. Telescopes have detected unusually high concentrations of the molecule phosphine–a stinky, flammable chemical typically associated with feces, farts and rotting microbial activity–in an atmospheric layer far above the planet's scorching surface.

The finding is curious because here on Earth, phosphine is essentially always associated with living creatures, either as a by-product of metabolic processes or of human technology such as industrial fumigants and methamphetamine labs. Although toxic to many organisms, the molecule has been singled out as a potentially unambiguous signature of life because it is so difficult to make through ordinary geological or atmospheric action.

Swathed in sulfuric acid clouds and possessing oppressive surface pressures and temperatures hot enough to melt lead, Venus is a hellish world. But the particular cloud layer where the phosphine is present happens to be relatively balmy, with ample sunlight and Earth-like atmospheric pressure and temperature. The results will have to be carefully vetted by the scientific community. Yet they seem likely to spark renewed interest in exploring our sister planet next door.

A Molecular Mystery

"It's a really puzzling discovery because phosphine doesn't fit in our conception of what kinds of chemicals should be in Venus's atmosphere," says Michael Wong, an astrobiologist at the University of Washington. Planetary scientist Sanjay Limaye of the University of Wisconsin–Madison agrees. "The bottom line is we don't know what's going on," he says. (Neither Wong nor Sanjay were involved in the work.)

After the sun and moon, Venus is the brightest object visible to the naked eye in Earth's sky. For thousands of years, people told stories about the glittering jewel that appeared around sunrise and sunset. Venus's brilliance is what made it attractive to Jane Greaves, a radio astronomer at Cardiff University in Wales. She typically focuses her attention on distant newborn planetary systems but wanted to test her molecular identification abilities on worlds within our cosmic backyard.

In 2017 Greaves observed Venus with the James Clerk Maxwell Telescope (JCMT) on Mauna Kea in Hawaii, searching for bar code–like patterns of lines in the planet's spectrum that would indicate the presence of different chemicals. While doing so, she noticed a line associated with phosphine. The data suggested the molecule was present at around 20 parts per billion in the planet's atmosphere, a concentration between 1,000 and a million times greater than that in Earth's atmosphere. "I was stunned," Greaves says.

Phosphine is a relatively simple molecule containing one phosphorus atom and three hydrogen atoms. It is known to reek of garlic or rotting fish, though by the time it reaches concentrations where humans can smell it, it is likely to cause lung damage. In the pilot episode of the series *Breaking Bad*, the character Walter White prepares phosphine gas to knock out two assailants who are threatening him.

Yet making the substance is not as easy as seen on TV. Phosphorus and hydrogen "hate each other," says Clara Sousa-Silva, a molecular astrophysicist at the Massachusetts Institute of Technology and a co-author of a study reporting the phosphine finding. "Hydrogen has much better stuff to do, and phosphorus would rather bond with oxygen. But if you throw enough energy at them, they can come together and be stable in some environments."

The gas giants Jupiter and Saturn contain phosphine because they have hot interiors where it can be energetically favorable to produce the molecule. Venus's runaway greenhouse atmosphere, by contrast, is full of oxygen-containing chemicals such as carbon dioxide that would normally soak up phosphine's phosphorus. For the molecule to be present at any level, let alone the amounts Greaves was seeing, was a genuine head-scratcher.

Meanwhile Sousa-Silva has built her career around studying phosphine—she goes by the handle @DrPhosphine on Twitter—predicting how it might appear in the atmosphere of a distant alien exoplanet. "I was considering these exotic worlds light-years away—super-Earths, tropical planets, sewage planets," she says. "And the whole time, it was just here next door."

The researchers and their colleagues made follow-up observations of Venus with the more powerful Atacama Large Millimeter/submillimeter Array (ALMA) in Chile last year, again detecting the atmospheric signature of phosphine. They then tried to come up with every possible reason for the strange molecule's existence, including volcanic activity, lightning strikes and even meteorites breaking up in the planet's atmosphere. "I think the best routes we could find fell short by a factor of about 10,000," Greaves says.

Of course, there might be additional pathways to making phosphine the team has not yet considered. But after exhausting their imaginations seeking out abiotic explanations, the researchers felt forced to acknowledge one other possibility in their paper, which appeared today in *Nature Astronomy*: the molecule could be made by life on Venus, just as life is the main way it manifests on Earth.

Life in the Clouds

Astrobiologists have long been enamored with Mars, a dry, rocky planet with conditions not all that dissimilar to those of Earth. More recently, they have become moonstruck by icy, potentially habitable worlds in the outer solar system, such as Saturn's geyser-spewing satellite Enceladus and Jupiter's oceanic moon Europa. But despite its drawbacks, Venus has not been entirely neglected by scientists speculating about life's extraterrestrial abodes.

From 50 to 60 kilometers above the Venusian surface, there is an atmospheric layer with pressure equal to that of sea level on Earth and temperatures between zero and 50 degrees Celsius. If not for the sulfuric acid clouds, one might call this layer "hospitable." Even so, there are terrestrial organisms that will happily tolerate such extremely acidic conditions in hot springs or other environments. This relatively clement region is precisely the place where the phosphine has been found.

Since the 1960s, astronomers have also noticed that Venus's clouds are not reflecting as much of the sun's ultraviolet light as they should be: an unknown *something* in the atmosphere seems to be preferentially absorbing that light instead. This observation led the late astrobiologists Harold Morowitz and Carl Sagan to propose that energy-hungry photosynthetic organisms might be the culprit. Meanwhile other researchers have never stopped searching for alternative abiotic explanations. Recent evidence suggests that the planet is still geologically active. And a model released earlier this year showed that Venus might have had an ocean for nearly three billion years–one that only disappeared a few hundred million years ago. Conceivably, life could have arisen on Venus when our sister world was much more Earth-like, only becoming airborne as the runaway greenhouse effect rendered the planet's surface uninhabitable.

"I've always thought it's as plausible to have life in the clouds of Venus as to find it in the subsurface of Mars," says David Grinspoon, an astrobiologist at the Planetary Science Institute, who was not involved with the study. "Each is an environment that could be habitable but isn't guaranteed to be."

Yet an almost equally good case can be made for Venus's clouds being inimical to life as we know it. Microbes have been found floating around in Earth's atmosphere, but none are known to exclusively spend their entire life cycle there. All of them have to land eventually, and Venus's surface seems too inhospitable a place to make for a good reservoir.

The Venusian area under consideration is also 50 times more arid than Chile's Atacama Desert, the driest place on our planet. And although it is true that living things have found good ways to thrive in aqueous environments tinged with traces of sulfuric acid, conditions on Earth's evil twin essentially reverse that formula: its cloud layer is mainly sulfuric acid with just a bit of water.

Venus Revisited

Venus remains an underexplored place. "Despite it being literally the planet next door, there are many mysteries that still need to be solved," Wong says. In order to rule out all nonliving explanations for the creation of phosphine, researchers will have to learn a great deal more about the planet itself, including its chemistry, geology and atmospheric physics, he adds.

Another issue might be the detection of phosphine itself. Noisy ripples that make resolving any particular line somewhat challenging are superimposed on Venus's spectrum in the team's data. These wavy structures could mimic a phosphine signature, says Bruno Bézard, a spectroscopist at the Paris Observatory. "I don't see a strong argument to say it's not a ripple," he says.

Greaves counters that the odds of finding the same signal using two separate facilities, JCMT and ALMA, is statistically small. Nevertheless, she and her colleagues are hoping to do additional observations at other wavelengths, such as infrared, to further test their initial results. Making higher-resolution maps of where the phosphine appears and seeing if it exhibits any kind of seasonal variation could also help tie it to biological processes.

In many ways, the unexpected finding appears analogous to the 1996 announcement of potential microscopic life in an ancient Martian meteorite designated Allan Hills 84001. Along with structures that looked like fossil bacteria, the sample contained an unusual form of iron crystals that appeared identical to those produced by microbial creatures on Earth. It took many years before researchers were able to figure out an inorganic explanation for those crystals.

Although life did not pan out as an explanation in that case, "it got everybody thinking, 'Well, why not?'" Grinspoon says. "Everything we know about Mars is consistent with that possibility. That led to a huge movement and catalyzed astrobiology as a field."

The phosphine finding might play a similar role in getting planetary scientists to pay more attention to Venus. In recent years, there has already been a contingent of researchers clamoring for more missions to our sister planet. Russia has proposed sending its Venera-D mission, which would include an orbiter and lander, to Venus as early as 2026. The European Space Agency similarly has the EnVision spacecraft on its drawing board, and it could reach its target in the next decade.

NASA is currently considering proposals for two different Venus missions for funding under its Discovery Program: the orbiting VERITAS and DAVINCI+. The latter would fly the first probe through Venus's atmosphere since the Soviet Vega balloons in 1985. A selection is expected sometime next year.

Any of these efforts, along with additional observations using telescopes on Earth, could help bolster or weaken the case for phosphine on Venus. Until then, many in the field are likely to reserve their full judgment. "It's very speculative to say that there is life on Venus," Wong says. "But it's also speculative to say there definitely can't be life on Venus."

For her part, Sousa-Silva is hoping the rest of the scientific community will subject her and her colleagues' methods and conclusions in the study to rigorous scrutiny. "I'm confident our models and data reduction are good, but I'm still skeptical," she says. "I expect the world to come and point out the mistakes I've made."

Such debates are important for science because similar conundrums are going to unfold every time someone claims evidence for life on a planet in our solar system or beyond, Sousa-Silva says. "I think it's very hard to prove something like this," she adds. "We have an innate desire to find life, and then we have our own rational minds that say, 'None of this is sufficient evidence.' We want to not be alone, but we also want to not be wrong. Sometimes those two things are hard to make coexist."

About the Author

Adam Mann is a journalist specializing in astronomy and physics. His work has appeared in National Geographic, *the* Wall Street Journal, Wired, *and elsewhere.*

Water Found in Sunlight and Shadow on the Moon

By Leonard David

For most of the space age, the moon has been considered a waterless world. In recent years, however, a steady drip-drip of discovery has shown that at least some parts of the moon—such as the large, permanently shadowed craters at its poles—contain significant deposits of water. This week, two new studies published in *Nature Astronomy* turn on the tap a bit more to the prospect of an unexpectedly watery moon.

The timing is good for NASA and other space agencies now planning ambitious human missions of lunar exploration and even settlement. After all, where there is water, there can be life—even if that life still requires space suits and radiation-hardened habitats.

A Possible Sky-High Signal of Sun-Warmed Water

The first new whiff of lunar water emerged from data gathered by NASA's Stratospheric Observatory for Infrared Astronomy (SOFIA). This modified Boeing 747SP jet provides its 2.7-meter telescope a view above 99 percent of the atmosphere's obscuring water vapor—a unique capability that allows agile observations in infrared without the use of space-based facilities.

In late August 2018 a team led by Casey Honniball, a NASA Postdoctoral Program fellow at the agency's Goddard Space Flight Center and a researcher at the University of Hawaii at Manoa, used infrared instruments onboard SOFIA to study the sunlit lunar surface. The observations, which spanned a mere 10 minutes, focused on a region at high southern latitudes near the moon's large crater Clavius, and they revealed a strong infrared emission at a wavelength of six microns (µm) from the crater and the surrounding landscape. Warmed by the sun, something on the lunar surface was reemitting the absorbed radiation just as molecular water—plain H_2O—would.

"We are unaware of any other material reasonable for the moon that exhibits a single spectral feature at 6 µm other than H_2O," Honniball and her fellow researchers report in their new paper. The authors suggest that the putative water is most likely stored in naturally occurring volcanic glass or sandwiched between microscopic grains of rock dust. Either scenario could provide shielding from the extreme temperatures and near-vacuum conditions on the moon's surface, allowing the water to persist. As to how it got there in the first place, no one is certain, but the leading explanation is that the water could have formed from free oxygen and hydrogen liberated from lunar rocks by micrometeorite impacts.

Using SOFIA is a new and unique approach for lunar science, Honniball says, but it is not the first time Earth-bound observations have revealed a six-micron emission from the moon. Balloon-borne observations by astronomers G. R. Hunt and J. W. Salisbury showed the spectral feature, she says. But Hunt and Salisbury made no mention of this in their paper on that research, published in 1969. Instead they focused on characterizing minerals on the lunar surface. "Maybe they just didn't know they made a huge discovery," Honniball speculates.

A Glass Half-Full

Honniball and her colleagues have already received additional time on SOFIA for follow-up observations. "We hope to map a majority of the moon to characterize the behavior of water," she says. "Does it vary across the lunar surface with lunar time of day and latitude? This will help us understand its sources and where it resides."

And that, in turn, could tell the world just how useful this newfound water might someday prove to be. Extraction will be straightforward if the water exists predominantly on the surfaces of rock grains: one will just need to scoop up lunar soil and subject it to moderate heating. If, however, the water is locked in glass, the material must be melted to release the water for collection—a much more energy-hungry process.

"Currently we do not have a good idea if the water we see with SOFIA is in amounts that make melting the glass worth it," Honniball says. "However, if we find abundances are high enough, this may be a more feasible option than mining water ice in permanently shadowed regions, which are extreme environments and hard to work in."

Jack Schmitt, a geologist who, as a member of the *Apollo 17* crew, remains the only professional scientist to have walked on the moon, says the SOFIA measurement may not be revealing true molecular water but something more fragile and transient. "The question that I would ask," Schmitt says, "is if the SOFIA data may be related to the possible weak bonding of solar wind hydrogen with oxygen at the surface of grains of silicate glasses and minerals in the regolith rather than being actual molecular water?"

One product of such reactions could be hydroxyl, a molecule just one hydrogen atom short of water. Honniball, however, says the six-micron emission seen by SOFIA is not consistent with hydroxyl.

Regardless of what substance is behind SOFIA's signal, Schmitt notes that basic chemistry should allow moisture to be wrung from even bone-dry lunar material. "Heating of hydrogen-bearing regolith to several hundred degrees would result in some of the hydrogen reacting with oxygen in silicates to produce water almost anywhere on the moon," he says.

Small Shadows, Immense Possibilities

Another paper published alongside the SOFIA study in *Nature Astronomy* spotlights an uptick in the distribution of permanently shadowed areas on the moon–sunlight-shy places known as cold traps–in which extremely low temperatures could freeze and sequester water essentially indefinitely, allowing it to accumulate into significant deposits over geologic time.

Scientists have studied such lunar regions for decades for their water-harboring potential, but previous work has focused on large cold traps within huge craters at the moon's poles. In contrast, this latest result extends the range of considered cold trap sizes down to one centimeter in diameter. Analyzing high-resolution imagery from NASA's Lunar Reconnaissance Orbiter, a team led by University of Colorado Boulder planetary scientist Paul Hayne found that such "micro" cold traps are far more prevalent than the well-studied large ones in the vicinity of the lunar poles. The new accounting raises the total surface area with the capacity to trap water to roughly 40,000 square kilometers–a pan-lunar region that, collectively, would be twice the size of Wales.

"The newly discovered micro cold traps are the most numerous on the moon, thousands of times more abundant than previously mapped cold traps," Hayne says. "If they are all full of ice, this could be a substantial quantity, perhaps more than a billion kilograms of water."

Hayne adds, however, that in situ sampling by robots or astronauts is required to properly assess their actual ice content. "What is really exciting about the micro cold traps is that they are much more accessible, which could enable more efficient extraction and utilization for both science and exploration purposes," he says. Indeed, this proliferation of tiny potential ice reservoirs could be much more accessible to future missions, Hayne says, because they exist in areas where a sunlight-bathed astronaut could comfortably and safely use a tool to reach into a dangerously cold shadow to dig out any ice.

For now, to further judge the value of micro cold traps, Hayne and his colleagues will use a high-tech camera dubbed the Lunar Compact Infrared Imaging System, which will voyage to the moon on the first south polar lander mission of NASA's Commercial Lunar Payload Services program as early as 2022. The camera will take close-up pictures of micro cold traps for the first time and will measure their temperatures.

Ground Truthing

On one hand, SOFIA and micro cold trap studies are welcomed news. Nevertheless, the big picture remains the same, says Ian Crawford, a lunar expert at Birkbeck, University of London.

Clearly, he says, the more easily accessible water there is on the moon, the greater the opportunities for its on-the-spot extraction and use to sustain immediate exploration efforts. Eventually the development of lunar water as a resource could spark an entire extraterrestrial economy in which the substance would become a lucrative feedstock for rocket fuel and other precious consumables. For now, though, "'ground truth' measurements are urgently required to confirm inferences made on the basis of remote-sensing measurements," Crawford says.

Angel Abbud-Madrid, director of the Colorado School of Mines' Center for Space Resources in Golden, Colo., also flags direct measurements as the most important next step to follow from the new findings. "What is now needed is to touch the lunar surface and gather detailed ground truth," he says. "Confirmation of not just the existence of water ice but its morphology, concentration, distribution and abundance is a must to proceed with existing exploration and resource-utilization plans."

About the Author

Leonard David is author of Moon Rush: The New Space Race *(National Geographic, 2019) and* Mars: Our Future on the Red Planet *(National Geographic, 2016). He has been reporting on the space industry for more than five decades.*

Asteroid Dust from Hayabusa2 Could Solve a Mystery of Planet Creation

By Jonathan O'Callaghan

The toaster-sized capsule hit our atmosphere at 12 kilometers per second, enduring temperatures of 3,000 degrees Celsius during its fiery descent before deploying a parachute to slow its velocity. It continued to fall until it finally reached terra firma in the Australian outback. Within hours teams of scientists located the capsule's landing site with radar and rushed via helicopter to retrieve it. Onboard were pieces of an asteroid, the biggest such haul in history, captured millions of kilometers from Earth and returned safely to our planet.

This event, the climax of the Japanese space agency JAXA's Hayabusa2 mission to the asteroid Ryugu, took place, in local time, on Sunday, December 6, 2020. It marked only the second time (the first being its predecessor, the Hayabusa mission, which launched in 2003) that a spacecraft has carried pieces of an asteroid back to Earth. With these samples, scientists hope to answer difficult questions about the history of our solar system and our own planet. How old are asteroids like Ryugu? How much water and organic material do they contain? And could they have first brought the raw ingredients for life to Earth billions of years ago?

While most groups of Ryugu researchers grapple with these outsized questions, a more rarefied cadre will be preoccupied with another, deceptively smaller matter: whether or not Hayabusa2's samples contain an intriguing ingredient of nearly all known meteorites. So far no one has been able to explain this ingredient's origins, yet the ramifications of doing so are tremendous. It may reveal to us not just some nebulous history of the solar system but also never before seen details of the process by which our sun's retinue of planets formed. In our understanding of how Earth—any planet in the cosmos, in fact—came to be, there may be nothing as important as the mystery of the chondrule.

Chondrules are small, seedlike rocks measuring up to just a few millimeters across and are thought to have formed some 4.5 billion years ago, shortly after the birth of our solar system. From there they became embedded in larger rocks, called chondrites, which make up the majority of the roughly 60,000 meteorites that humans have discovered throughout recorded history.

"Chondrules are everywhere," says Fred Ciesla, a planetary scientist at the University of Chicago. Even so, scientists have been unable to agree on how they formed for almost two centuries. Some think they were by-products of planet formation; others posit they were the seeds of planet formation itself. Either way, the menu of chondrule creation scenarios is vast, ranging from lightning-fused dust to colliding chunks of protoplanets to giant, gas-heating shock waves rippling through the primordial cloud of material that surrounded our newborn sun.

Understanding chondrule formation could, in other words, reveal our solar system's earliest moments. And now, with fresh or prospective results from missions such as Hayabusa2 as well as other avenues of research, chondrule-obsessed scientists are on the cusp of answering the long-standing question of where they–and perhaps we–came from. "They are stained-glass windows to the earliest time period of the solar system," says Harold Connolly, a cosmochemist and chondrule expert at Rowan University. "They are witnesses to processes that operated in the early solar system. The question is, What did they witness?"

Droplets of Fire

In 1802 British chemist Edward Howard was one of the first scientists to recognize chondrules as "rounded globules" in meteorites. Their name, given later by German mineralogist Gustav Rose and Austrian mineralogist Gustav Tschermak, originates from the Greek *chóndros* ("grains") and the German *kleine kugeln* ("small balls"). In 1877 British scientist Henry Sorby would characterize them in greater detail, describing chondrules as "droplets of fiery rain," molten globules that condensed around the sun, although then, as now, no one knew exactly how they formed.

The broad outlines of our solar system's genesis are clearer. This creation story, which scientists have assembled through decades of observation and modeling, begins more than 4.5 billion years ago, when dust and gas from a giant molecular cloud gravitationally collapsed to create a protostar that would become our sun. This protosun was surrounded by a spinning disk of gas and dust. Within this disk, a mix of gravity, aerodynamics and electrostatic force caused grains of dust to stick together, forming larger and larger agglomerations–such as planetesimals, the kilometer-scale building blocks of planets–and within a few million years the planetesimals coalesced into planets. These worlds gradually settled into the familiar forms and orbits we know today. But if the outlines of this story are clear, the details remain mysterious. Chondrules appear in the opening chapters, somewhere amid the leap from dust to planetesimals. How do you get from microscopic motes to entire worlds thousands of kilometers across?

Chondrules are essentially rocks within rocks. They appear as rounded flecks in chondritic meteorites. Some are visible to the naked eye, whereas others can be seen only under a microscope. It is difficult to overstate just how abundant chondrules are: Despite the fact that none are known to have survived the process of incorporation into planets, they are very common off-world, often constituting the bulk of material within chondrite meteorites. Some chondrites are so packed with chondrules they look almost like a conglomeration of beads.

Made of minerals such as olivine and pyroxene and sometimes glass, chondrules themselves come in a variety of shapes, sizes and compositions. They often contain a glittering array of crystals. Scientists can date their formation to a window of a few million years, circa 4.567 billion years ago, by measuring the abundance of aluminum 26, a short-lived radioactive isotope they contain. That chronology makes chondrules the second-oldest recognizable objects in our solar system, after calcium-aluminum-rich inclusions (CAIs), specks of white in meteorites that are thought to have formed one million to three million years earlier by condensing out of the gas that surrounded our young sun.

There are many classes of chondrites. Ordinary chondrites, for example, are full of chondrules and account for more than nine out of every 10 chondrule-containing space rocks. Carbonaceous chondrites, which account for about 4 percent of all chondrites, typically have a high abundance of carbon; the most carbon-rich ones are thought to have formed in the outer solar system. A subgroup called CI chondrites possesses only microscopic chondrules because larger ones were weathered away by liquid water that once flowed through the parent body. And CB chondrites hold the distinction of being the only type of chondrite for which there is near-universal agreement on how they formed. "These are a group that we think formed during one giant impact," says Sara Russell, a planetary scientist at the Natural History Museum in London. That pinned-down provenance makes them "kind of magical."

As for the exact origins of all the other chondrite varieties? Those remain anyone's guess. "It's frustrating, but it's also kind of fun that we don't know," Russell says. "They're obviously telling us about some ubiquitous, super important process about how our solar system formed. We just have to work out what that is."

An Impossible Problem

In 2000, at the Lunar and Planetary Science Conference in Houston, a stunned audience watched as John A. Wood, then at Harvard University and one of the most revered scientists in meteoritics, appeared to admit defeat in understanding the origin of chondrules. Like many before him, Wood became fascinated by chondrules at first sight. "[These] little balls of stone were so charming and interesting and mysterious that I just got seduced by them," he says. But he was frustrated at the lack of progress that had been made. "We still don't understand what the meteorites are telling us, and sometimes I wonder if we ever will," he wrote in a summary of his speech. A few years later, facing a lack of funding, he opted to retire, turning his attentions to oil painting and spending time with his wife. "I quit science cold turkey," Wood says.

The speech was a shock to many. "He basically said he'd wasted his entire life working on chondrules because it was an impossible problem to solve," says Conel Alexander, a cosmochemist at the Carnegie Institution for Science in Washington, D.C. "That got a lot of people pretty upset." Larry Nittler, also a cosmochemist at Carnegie, who was in the audience, says he "actually stood up" to defend chondrule research. "I said, 'I'm still excited about these incredible rocks from space,'" he recalls. "I don't think I've gotten as much attention or praise for anything I've done in my career. The whole room erupted in applause."

Wood's pessimism is understandable. After all, space scientists have managed to definitively solve a wealth of seemingly intractable mysteries. They have teased apart the first moments of the universe's existence, discovered worlds around other stars, observed gravitational waves and captured images of a black hole. Against such achievements, the stubborn enigma of the lowly chondrule seems to shrink even smaller than its already niche status. Today there are, the joke goes, as many theories about chondrule formation as there are chondrule scientists themselves–and tomorrow there will inevitably be more.

The problem of chondrules has from the start been intergenerational, inspiring one cohort after another to try tackling the issue, with varying success. The main problem is finding a model that can explain all the different, diverse properties of chondrules. "There are no models that tick all the boxes," Alexander says. For chondrules to form, dust must have been heated to temperatures of up to 2,000 degrees C by some process in the early solar system, before rapidly cooling over just days or even hours. This process, whatever it was, most likely occurred throughout the solar system; that seems to be the only way to account for the large abundance of chondrules found in chondrites on Earth. The telltale rings of accumulated dust found in the centers of chondrules suggest that they must also have drifted for a time through the dusty environs around our newly coalescing sun.

Most chondrule scientists fall into one of two camps. The first believes chondrules were among the earliest solid objects to appear in the solar system, forming directly from the solar nebula–the cloud of dust and gas that surrounded our young sun. This would make chondrules a key stepping-stone from minuscule dust to larger kilometer-sized planetesimals. The second camp believes that chondrules were not among the first solids to form but actually arose after planetesimals–perhaps even after the planets themselves. They were, in this view, by-products of the planet formation process rather than an active part of it.

Within the first camp, one idea is that gravitational instabilities in the disk of dust and gas around our sun resulted in "shock fronts" that fused some of the dust into chondrules. "If you look at a picture of a galaxy, you see spiral arms spiraling round, and the same thing would have happened in the protoplanetary disk," says Rhian Jones, a cosmochemist at the University of Manchester in England. "And you can produce shock fronts associated with those density differences between the clumpy arms and the gas."

The more radical nebula-lightning model, meanwhile, proposes that friction between dust and gas particles around the sun sparked immense discharges of electricity that fused this dust into chondrules–although it is not clear how such lightning would be produced. Some models propose chondrule creation factories emerging from sheets of electric current trapped by huge magnetic fields within the spinning protoplanetary disk. Such "hotspots" could have been tens or hundreds of thousands of kilometers across and would have melted dust grains to churn out the primordial globules.

In the other camp, whose members argue that chondrules formed after the planetesimals, one of the more prominent models is called impact jetting. Here planetesimals would collide at high velocities, creating the necessary heat to produce chondrules. "It essentially squirts out some molten material that could break up into droplets," says Brandon Johnson, a planetary scientist at Purdue University. A variant of this, called splashing, would have involved collisions between molten objects at lower velocities, releasing droplets into space that solidified into chondrules.

The nebula-shock model, meanwhile, posits that Mars-sized planetary embryos moving through the nebula could act like boats sailing through water, fusing dust into chondrules. "As it's moving through the gas at supersonic speeds, it drives a bow shock," says Steven Desch, an astrophysicist at Arizona State University. "The chondrule precursors, this dust, are heated by entering this compressed hot gas and are processed by it."

Other ideas include radiative heating, a relatively new inclusion in some models that suggests planetesimals flying low over molten bodies could have been roasted and then cooled to produce chondrules. Rocks that underwent this natural "heat treatment" would be sturdier and more likely to survive the passage through Earth's atmosphere, which would explain why most of the meteorites we find are chondrites. "The meteorites got hardened, and Earth's atmosphere is a filter that is weighted toward really dense and hard stuff," says astronomer William Herbst of Wesleyan University, one of the researchers behind the idea.

Against this rising theoretical tide, some wilder notions have already been ruled out. Events from outside the solar system such as gamma-ray bursts—enormously energetic explosions from sources such as merging neutron stars or black holes—were once considered a possibility but now seem implausible because of the great distances involved. Even so, many models still remain, complicated by the fact that chondrules are not really predicted by planet formation at all. "We can build a story about how planets form without ever invoking chondrule formation," Ciesla says. "It's obvious that there's a piece of the story that we're missing."

Narrowing down which of the remaining theories is correct is hard, and arguments can get heated. "The best way to make friends and enemies in meteoritics is to publish another chondrule-forming model," Connolly says. At stake is what role chondrules played in our solar system. If they were among the first solids to form, then some inescapable process took place around our young sun that could explain how planet formation begins around most any star. But if not, are they less vital to the process than once thought?

"I'm putting my money on collisions right now," says planetary scientist Eugene Chiang of the University of California, Berkeley. "And to be totally up front about it, that makes [them] a little less interesting. Because if you're interested in planet formation, it means the chondrules are not the most primitive objects. They're secondary products."

Cooking up Chondrules

Most of our ideas on chondrule formation come from modeling the early solar system and performing experiments on Earth to replicate different formation methods. Meteorite scientist Aimee Smith of the University of Manchester and her colleagues are one of several teams around the world that perform such experiments, mixing chemicals into a powder to resemble different types of known chondrule compositions. Then they place the powder in a furnace and heat it to extremely high temperatures for anywhere from hours to days, before cooling it to mimic different formation models. "If we get ones that look similar to the natural chondrules that we've studied, then we have a better idea of how they formed," Smith says.

Experiments such as these are designed to work in concert with solar system modeling. "The experiments are just defining the conditions for chondrules," says Jones, Smith's collaborator. "But models are trying to come up with scenarios in which those conditions are satisfied." And in our solar system, such modeling is starting to paint a new picture of its earliest moments.

Recent work on measurements of isotope ratios in meteorites indicates that two different reservoirs of chondrites formed early on—one in the inner solar system and one in the outer solar system, where chondrules may have been produced separately. These separate populations would have mixed together after Jupiter, having initially formed more than twice as close to the sun, migrated out to its present position, an idea called the grand tack hypothesis. If true, this would suggest that the story of our solar system's turbulent history is stored within chondrules themselves, offering yet another reason to lavish them with careful attention.

Elsewhere, observations of other solar systems–in particular, protoplanetary disks of dust and gas around young stars–are yielding information about possible scenarios for chondrule creation. In 2014 astrophysicist Huan Meng, then a graduate student at the University of Arizona, and his colleagues reported a flash of infrared light around a star called NGC-2547 ID8 over 1,000 light-years from Earth–evidence for a potential protoplanetary smash-up. Though not definitively linked to the formation of chondrules, the observation at least showed that suitably energetic collisions to make them do seem to occur in young systems. "Before [our] paper, we didn't have any direct hard evidence of any extrasolar planetary impacts," Meng says.

In the future, astronomers should also be able to probe the distribution of dust around young stars with higher-resolution images, which could make it possible to refine some models of chondrule formation. "With improvements in techniques and telescopes, now we can start to see the dust production around young stars," says Yves Marrocchi, a planetary scientist at the French National Center for Scientific Research. "Maybe in the near future we can see the formation [process] of chondrules." Such telescopes may include NASA's much delayed James Webb Space Telescope, scheduled to launch this October.

If chondrules are among the first solids to form around stars, they may be crucial catalysts for subsequent planet formation, in particular the jump from dust-sized to kilometer-sized objects. "There's a gap when you need multimeter- to kilometer-size objects to actually form the rocky parent bodies of planets," Connolly says. "What happens in that gap becomes really important." And perhaps chondrules are even more essential; one model of planet formation, pebble accretion, posits that larger bodies swept up pebble-like dust to grow into planets. Could those pebbles in fact be chondrules? "We don't know if they're the same," says André Izidoro, a planetary scientist at Rice University. Finding out would likely require "a big sample of an asteroid," he says. And as it happens, we just got one.

"No Single-Sentence Answer"

Early assessments revealed that Hayabusa2 managed to bring back more than five grams of material from asteroid Ryugu. According to Shogo Tachibana, who leads JAXA's sample-analysis team, that should be more than enough to see if chondrules are present. He and his team began studying the samples early this year, after they were transported from Australia back to Japan. Most of their results are still forthcoming. "We don't know if chondrules in Ryugu are different from other types of chondrules in other chondrites," Tachibana says. Ryugu appears to be similar to Earth's carbonaceous chondrites, so most experts expect chondrules to be present in the samples, but as of this writing, no one yet knows whether they will resemble those already in collections or will be like nothing ever seen before.

It is possible that Hayabusa2's samples do not contain chondrules at all. "I think that would be shocking to the chondrule community," Herbst says. "If there were no chondrules, and it looked like there had never been chondrules in them, then maybe chondrule formation is not such a ubiquitous process," Russell says.

Early results from a lander called MASCOT, deployed by Hayabusa2 onto Ryugu in October 2018, have already tantalized scientists. Images from the lander showed many white markings on the surface, which may have been CAIs but also could have been chondrules. "We were surprised that we really could see [the markings] and that there were so many of them," says Ralf Jaumann, head of MASCOT's science team at the German Aerospace Center, or DLR, Germany's space agency. Only chemical studies of the samples conducted back on Earth will reveal the nature of those markings.

If there are chondrules in the Hayabyusa2 samples and if they are similar to chondrules scientists have already studied, it will be possible to pinpoint the location, time and perhaps even conditions in which they formed. If the samples contain new types of chondrules, however, that could provide a fresh perspective on the larger problem of the origin of the solar system. Scientists such as Connolly would welcome such a scenario. "I certainly hope there are a few surprises and we find objects we didn't expect," he says. And even if chondrules are not present, that could simply suggest that water made liquid from the heat released by radioactive decay, impacts and other sources had long ago erased evidence of its chondrules, similar to the origins of CI chondrites found on Earth.

Hayabusa2 is not the only sample-return mission with extraterrestrial gifts in store for chondrule scientists. NASA's OSIRIS-REx spacecraft is scheduled to return to Earth in September 2023 with recently acquired samples of another asteroid, called Bennu, that are expected to be chondrule-rich. "It would be really disappointing if we didn't find chondrules in the material," says Connolly, who is also part of the OSIRIS-REx team. "I'm looking forward to finding chondrules that I know and chondrules that I don't know."

If researchers ever manage to definitively determine how chondrules formed, that could go a long way toward revealing whether or not they were crucial to the subsequent creation of Earth and our sun's other small worlds. Presuming, of course, that the creation story ultimately revealed is relatively straightforward. Some experts, however, suspect no simple solutions will be found, in part because more than one theory is correct. "I do not think it is a single-sentence answer," says planetary scientist Sarah Stewart of the University of California, Davis. "There were probably many droplets being made in different ways." Russell agrees: "My favorite theory is that everyone's right. All these processes happened somewhere in the solar system. There were shock waves, there were impacts, there were bow waves, there was lightning. I think these things all happened, and they all formed chondrulelike objects."

Which may mean Wood was on the right track all along when he made his infamous, career-capping declaration of futility: If nearly every idea for chondrule creation reflects a process that actually occurred in the solar system's ancient history, there may be no deeply meaningful distinctions among them. But that possibility won't keep new generations from trying, just as their predecessors did. "If I had to do it all over again, I would have made the same attempt," Wood says. And to anyone following in his footsteps? "I would wish them good luck."

About the Author

Jonathan O'Callaghan is an award-winning freelance journalist covering astronomy, astrophysics, commercial spaceflight and space exploration. Follow him on X (formerly Twitter) @Astro_Jonny.

Excitement Builds for the Possibility of Life on Enceladus

By Annie Sneed

Saturn's frozen moon Enceladus is a tantalizing world—many scientists are increasingly convinced it may be the best place in our solar system to search for life. NASA's Cassini spacecraft, currently orbiting Saturn, has made intriguing observations of icy jets spewing from a suspected underground liquid ocean on the mysterious world that might be hospitable to alien life.

Cassini's tour is due to wind down in 2017, and scientists badly want to send a dedicated mission to Enceladus to look for signs of life. In fact, some have already started seriously thinking about exactly how they might do this—including planetary scientist Carolyn Porco, who is the imaging team leader for Cassini. Earlier this month, she gathered a group of researchers including oceanographers, organic chemists and astrobiologists at the University of California, Berkeley, to strategize how to search for extraterrestrials on Enceladus—which, according to Porco, "is a total b*tch of a problem to solve."

Although Enceladus is small in size and shrouded in a thick shell of ice, it appears to be a habitable world: It has a source of energy from friction created by its orbit around Saturn, organic compounds that are building blocks for life and a liquid water ocean underneath all that ice. But just because Enceladus may be hospitable to life does not mean life exists there; it will take much more work to definitively prove it. At the Berkeley meeting, scientists laid out the data Cassini has collected for Enceladus—they discussed analyses of its geysers, measurements of its ice shell, ideas on what its ocean chemistry might be like, and more. Yet even with all the newest data and models scientists have, they are not even close to detecting organisms on Enceladus—hence the need for a space mission.

Finding life there would be a profound revelation that we are not alone in the cosmos. Furthermore, the discovery of organisms—or the lack thereof–could answer the subtler mystery of how life started on Earth. Researchers at the meeting presented two major opposing theories about how life here originated (in the ocean versus on land), and the group discussed how exploring Enceladus would inform this debate. "It would be a test of one of the ideas about the origin of life," Porco says–specifically, the proposition that Earth's species sprang in the sea. For example, if organisms exist in Enceladus's ocean and presumably arose there, it would support the theory that life began on Earth in hydrothermal vents (hot, nutrient-rich, deep-sea vents on the ocean floor) rather than in patches of water on land.

Enceladus could also teach us about genesis in our solar system in other critical ways. "You're not just searching for life, you're searching for an understanding of the nature of that life, and how it compares to life on Earth," says Chris McKay, a planetary scientist at NASA Ames Research Center. For instance, if we discover that creatures on Enceladus are nothing like those on Earth–if their biochemistry is completely different–then it would likely mean that the two forms of life arose separately and independently, and thus, that aliens might be likely to exist other places as well. "If life started at least twice in our solar system, then you know the universe is full of life," McKay says. Or, if we find out that Enceladus organisms and Earth organisms are made in identical ways, it may indicate that life originated someplace else, and was carried to both worlds. If Enceladus is barren, however, it could support the theory that life needs an environment on dry land to get started, not an ocean. Regardless of what a mission to Enceladus might discover, the answer will tell us something fascinating.

Skimming Geysers

Enceladus has over 90 geysers that spew plumes of salty water vapor, organic compounds and ice particles from the underground ocean into the air. These present a great opportunity for a visiting spacecraft, which would not have to land to search for life (which is much more difficult and expensive) but could simply fly through the geysers to capture samples. "The plume is coming right out of the ocean," McKay explains, "So why would we want to land? We can get the freshest stuff, coming right from the source."

Yet even if life exists on Enceladus, it may or may not show up in plume samples. If the pelagic ocean on Earth (that is, the open water away from the shore or seafloor) is an analogue for Saturn's icy ocean moon, then the outlook is depressing–the pelagic zone has an extremely low density of life even on our planet. "If we had this in Enceladus's ocean, it would be very hard to even pick up an organism," Porco says. Scientists would need to sample a ridiculously large amount of water in order to capture any organisms.

Thankfully, a few months ago a microbiologist told Porco about decades-old scientific research that makes her optimistic about finding life in the plumes. At the Berkeley meeting, she described this research on a process called "bubble scrubbing" that occurs in Earth's oceans–and it could make quite a difference in Enceladus's geysers. It turns out that wherever bubbles rise through water, they scrub the water column so that organisms and organic materials become concentrated at the surface. And when the bubbles burst (like in ocean spray or in Enceladus's jets), they eject those microbes in the spray. So if life exists on Enceladus, its plumes may contain a much greater concentration of organisms than the rest of its ocean–all thanks to bubbles. "Even if the ocean on Enceladus starts out being as microbially poor as the pelagic ocean on Earth, which is the worst case, we still have a chance of seeing lots of organisms in the plumes," Porco says. Still, this scenario immediately presents another issue: A spacecraft must find a way to capture a sample without smashing the delicate organisms to bits as it makes a high-speed pass through the jets.

Seeking Signs of Life

Once a spacecraft collects a sample from Enceladus, how will scientists test it for life? The process is more complex than simply searching for something that is alive–after all, researchers have argued over the definition of life for years. In the case of hunting for extraterrestrials scientists must get creative. "If you went to Mars and found a dead rabbit on the ground, it's not alive, but it is compelling evidence of life," McKay says. "So we're not searching for something that's alive but searching for the molecules that life uses. In other words, we're looking for the body of the dead rabbit."

The molecules that McKay and other scientists consider most important are amino acids–the building blocks for proteins. "They occur on comets and meteorites, so if there's a primordial soup on Enceladus, it should have amino acids," McKay says. "They're so incredibly useful and so good in water that life would be pretty dumb not to use them." At one point in the Berkeley meeting, however, Porco brought up a critical point: What if Enceladus organisms aren't made of amino acids? McKay replied, jokingly, "Then we're sunk and nature is perverse. We should all just give up and become poets." What he meant is there's a large consensus in the scientific community that amino acids will be useful in the hunt for life–and if such thinking turns out to be wrong, well "then we're even dumber than we thought we were," McKay says.

Another important signature scientists want to detect is lipids, which cells use to build their outside walls. "It's a similar story to amino acids," explains Alfonso Davila, a research scientist at the SETI Institute and Ames. "They're something you'd expect to be present at the origin of life and you'd expect cells to use them." Scientists will need to do more than simply detect amino acids and proteins on Enceladus–both of those molecules exist on their own in many environments, with or without life. But astrobiologists can target distinct structures and distributions of amino acids and lipids they think are unique to life. "We're looking for molecules and structures that life makes, that are distinctly different from the random mess that chemistry makes," McKay notes.

Some other possibilities on the search list are large organic compounds as well as photos of actual organisms taken by the spacecraft of samples from the plume. That might be, for example, images of an organism swimming or eating. Such a find might be the most direct evidence of something "alive" but many researchers have doubts about the plausibility of imaging organisms, Davila says. "It's one of those high-risk, high-reward experiments. The likelihood of a negative result is very high," he adds. "It's very hard to tell the difference between a cell and a dot that is just a particle." This issue was hotly debated at the meeting but some scientists sounded more hopeful–they discussed new techniques they are exploring to reliably image Enceladus's microbes (if they exist).

Ultimately, scientists think it will probably take a combination of evidence to show that they have actually found life. And, of course, cost and technology will constrain the experiments they are able to perform. The hunt will be incredibly complicated, especially considering that organisms on Enceladus may not look or operate anything like they do on Earth. "We're walking a thin line between what we know based on Earth life and what we expect life would be like otherwise," Davila says. "It's one of the things that prevents us from coming up with a good strategy."

Or, to reiterate Porco's observation–"It's a total b*tch of a problem to solve."

About the Author

Annie Sneed is a science journalist who has written for the New York Times, Wired, *Public Radio International and* Fast Company.

Pluto's Wonders Come into Focus

By Lee Billings

Last July, NASA's New Horizons spacecraft flew by Pluto, the last unvisited world of the classical solar system. As the largest known member of the Kuiper belt, Pluto is also the gateway to a new frontier, a scarcely studied collection of primordial icy bodies far from the sun that constitutes the "third zone" of the solar system after the realms of the inner rocky planets and the outer gas giants.

Like most first glimpses of new frontiers, Pluto held so many surprises for New Horizons that the past eight months have seen a steady stream of discoveries coming from the mission, as the spacecraft's small radio transmitter beams its gathered data back home. The biggest surprises have been Pluto's surface and atmosphere, which are restlessly active and diverse despite average temperatures of only tens of degrees above absolute zero. Some scientists expected New Horizons would find Pluto to be little more than an inert, sunlight-starved orb. Instead, the spacecraft encountered a world where nitrogen glaciers flow down into plains of frozen methane from towering mountains of water ice. Sunless half-frozen oceans lurk deep beneath the surface, and multiple moons tumble overhead through hydrocarbon-hazed red skies that tinge to blue at sunrise and sunset.

But beyond celebrating the visceral thrill of the Pluto flyby itself, or the intellectual frisson of gazing on full-color close-ups of a place so alien and faraway, most of these discoveries from New Horizons have so far found a muted public reception. The story has simply been that we went to Pluto, and witnessed wonders. What those wonders actually mean—for our understanding of Pluto, for planetary evolution, and for the broad history of the solar system—is something that the mission scientists themselves are still working out. They summarize their latest thoughts in this week's edition of the journal *Science*, with a quintet of papers that constitute the synthesis of our current understanding of Pluto.

Here are the three big-picture takeaways from our emerging portrait of this strange, frozen world:

1. Pluto Is Young at Heart

New Horizons was only able to closely study one hemisphere of Pluto as it whizzed by, revealing a sprawling heart-shaped plain of mixed nitrogen, carbon monoxide and methane ices ringed by mountains and heavily cratered terrain. Dubbed Sputnik Planum, the 1,000-kilometer-wide western lobe of the heart looks almost bubbly, like a churning pot of creamy oatmeal or frothy foam on a pint of Guinness beer.

Sputnik Planum has no craters, and is likely less than 10 million years old, probably formed from fresh snow and glaciers sliding down from nearby rugged highlands. Its bubbles are convection cells driven by heat rising through the thick ice from deep below. A small "mountain range" at Sputnik Planum's northwestern edge is in fact blocks of water ice that seem to be bobbing in the higher-density ices like cubes in a glass. These blocks are perhaps crust fractured and overturned by some tectonic upheaval. To the south, New Horizons scientists have spied what seem to be two young cryovolcanoes, Wright Mons and Piccard Mons, relatively unblemished kilometers-high mounds surrounding central pits at least as deep.

Taken together, these features show that more than four billion years after its formation, Pluto still somehow retains enough internal heat to maintain an active geology and, here and there, a very youthful surface replenished by cryovolcanism and the seasonal sublimation and deposition of volatile ices. Deep within the world, Pluto's heat could be sufficient to sustain an ocean of ammonia-rich water beneath a thick roof of water-ice bedrock. Long linear striations upon parts of Pluto's surface hint that any subsurface ocean may be slowly freezing, deforming the ground and releasing additional latent heat as it turns to ice.

2. Most of Its Surface Is Still Ancient and *Really* Weird

Sputnik Planum's smooth-featured youth is exceptional. Most of the rest of Pluto's exterior is far more craggy and ancient, altered extensively across hundreds of millions or billions of years. Varying mixtures and combinations of nitrogen, water, carbon monoxide and methane that make up Pluto's crust create different varieties of ice and terrain, similar to how rocks on earth can form cliffs of soft chalk or mountains of hard granite. These varying substrates can then be textured with pits, grooves and channels produced by subliming ice, eroding glaciers and precipitating frost–effects driven by Pluto's weather, which fluctuates in decades-long seasons.

The results are usually bizarre, and difficult to decipher. Northeast of Sputnik Planum, past stretches of pitted plains, the surface is wrinkled with closely spaced ridges that rise sharp and knife-like half a kilometer into the air. This "bladed terrain" may be vestiges of an old, once-buried layer of highly durable material exhumed and weathered by some combination of scouring ice, swirling winds and glaring sunlight. Or it may be newer, formed from airborne methane frost glazing rigid crests of water ice. What is clear is that Pluto's landscape cannot be understood without also closely studying its weather–its atmosphere.

New Horizons has revealed Pluto's tenuous atmosphere of gaseous nitrogen and methane to be colder and more compact than previously thought, and layered with hazes of soot-like hydrocarbon particles produced by ultraviolet light and cosmic rays. The particles are reddish, but at sunrise and sunset when sunlight passes through the thickest hazes, they scatter the light to give Pluto's sky a blue tint. The particles are also sticky, and grow like snowflakes over tens of thousands of years, until at last they become heavy enough to fall, accumulating as crimson sludge in the world's most ancient terrains.

The most primordial part of Pluto's surface may be a hemisphere-spanning splash of red called Cthulhu Regio, a region so thoroughly pulverized by craters it is thought to be some four billion years old.

Curiously, it is directly adjacent to the western edge of what could be Pluto's youngest landform, the fresh, cream-colored ices of Sputnik Planum. Even Sputnik Planum, it turns out, has surprisingly ancient roots: Its youthful ice fills a deep basin that may be the oldest, largest impact crater still in existence on Pluto.

3. Pluto's Moons Are Chips Off the Old Block

Besides Cthulhu Regio, Pluto's most notable other old, reddish, impact-generated feature isn't actually on the dwarf planet at all—it's Pluto's largest moon, Charon. Long thought the product of a cataclysmic impact of the same sort that made Earth's moon, Charon's violent origins have been all but confirmed by New Horizons. Most of Charon's surface is actually grayish bright water ice, with craters indicating it is more than four billion years old—a strong hint it coalesced from shattered and ejected pieces of Pluto's water-ice crust. Its connection to Pluto hasn't been completely severed, though: Mordor Macula, a cap of dark red hydrocarbons at its north pole, is likely produced by ultraviolet light reacting with wisps of upper atmosphere that drift away from Pluto's gravity and freeze onto Charon, building up like layers of red varnish over billions of years.

Much like Pluto, Charon also seems to have a subsurface ocean—or at least it used to. New Horizons discovered a deep gash snaking across some 1,800 kilometers of the moon's surface—a furrow four times longer than the Grand Canyon on a moon roughly the size of Texas. The gash seems to be from a time when Charon's inner ocean froze, bulging as it turned to ice and rupturing the moon's crust from within.

In addition to Charon, Pluto has four much smaller moons: Styx, Nix, Kerberos and Hydra. New Horizons has found them to be much brighter and smaller than most researchers expected. They all are spinning rapidly, and have extreme axial tilts so off-kilter to those of Pluto and Charon that they are not easily explained. These smaller moons were probably also produced by the great Charon-forming impact. Like Charon, they seem to possess crater-battered surfaces of four-billion-year-old water ice, and they are oblong rather than spherical, as if they are all less moons and more barely-held-together piles of coalesced rubble. At least one of them, Kerberos, is shaped like a dumbbell, signaling its formation from two smaller bodies merging after the cataclysmic impact.

What's next for all these weird worlds of staggering geological richness? We'll find out soon. Fully half the data New Horizons took remains onboard, still awaiting transmission. Meanwhile, the spacecraft is continuing its mission, cruising toward a rendezvous with a smaller, more distant Kuiper belt object in 2019. The best may be yet to come.

About the Author

Lee Billings is a science journalist specializing in astronomy, physics, planetary science, and spaceflight, and is a senior editor at Scientific American. *He is the author of a critically acclaimed book,* Five Billion Years of Solitude: the Search for Life Among the Stars, *which in 2014 won a Science Communication Award from the American Institute of Physics. In addition to his work for* Scientific American, *Billings's writing has appeared in the* New York Times, *the* Wall Street Journal, *the* Boston Globe, Wired, New Scientist, Popular Science, *and many other publications. A dynamic public speaker, Billings has given invited talks for NASA's Jet Propulsion Laboratory and Google, and has served as M.C. for events held by* National Geographic, *the Breakthrough Prize Foundation, Pioneer Works, and various other organizations. Billings joined* Scientific American *in 2014, and previously worked as a staff editor at* SEED *magazine. He holds a B.A. in journalism from the University of Minnesota.*

Section 2: Hunting Exoplanets

Feeling Hot, Hot, Hot: Astronomers Discover a Giant Planet Hotter Than Most Stars

By Leslie Nemo

After a decade of searching and verification, a research team revealed a new exoplanet, and you could say it's coming in hot. The orb is the hottest giant exoplanet ever found, according to a study published today in *Nature*. "KELT-9b" intrigues astronomers not just because of the unusual chemistry the heat is bound to create in its atmosphere, but also because of the opportunities it offers for more in-depth studies.

"We're fascinated with the weirdness that nature hands us," says B. Scott Gaudi, an astronomer at the Ohio State University who helmed the study, as he explains why his academic circle is ablaze with KELT-9b curiosity. First and foremost, there is the planet's impressive heat. Gaudi and his team calculated KELT-9b to be roughly 3,777 degrees Celsius (6,830 degrees Fahrenheit) on the dark side and 4327 degrees C (7,820 degrees F) on the star-exposed side. This makes KELT-9b almost 20 percent hotter than the next-most-scorching exoplanet, WASP-33b. The new planet's blistering temperature is thanks to its parent star, KELT-9. The star falls in the A-type bracket of stars, the third-hottest category. It is the seventh star of its classification to be found with an exoplanet, and at roughly 300 million years old, it is particularly young and toasty–9,897 degrees C (17,846 degrees F), to be exact. That is nearly twice the temperature of our own sun.

With extreme conditions come extreme attributes. KELT-9b receives 700 times as much solar radiation as WASP-33b. That amount is similar to the glow of an entire K-type star, the second coolest type, and is more than enough to break down molecules to their constituent atoms. Gaudi predicts, then, that the atmosphere of KELT-9b is writhing with ions in a star-like pandemonium. Drake

Deming, an astronomer at the University of Maryland, agrees. "KELT-9b is a hybrid planet-star. It has the mass of a planet, and [yet] an atmosphere like a star," he says.

The group's finding could also influence astronomy research in a slightly less splashy but equally important way. "By and large, the population of planets around higher-mass stars has been relatively unexplored," Gaudi says. Early exoplanet discoveries happened around low-mass stars. These findings inspired a trend of researchers pursuing stars of lower and lower masses. In an effort to carve out their own niche in the KELT, or Kilodegree Extremely Little Telescope, project, Gaudi and his team targeted higher-mass stars. The discovery of KELT-9b might encourage more researchers to dig into planets on the opposite end of the typically researched spectrum, Gaudi explains. Deming also thinks the find has opened the door for a new approach to exoplanets. "This is a new class of planet," says Deming. "I think we'll find a lot more of these."

Even if the hunt for companions to massive star grows, researchers will continue their study of KELT-9b. The planet is a good candidate for high-quality, precise characterization by a variety of telescopes, says Adam Burrows, an astrophysicist at Princeton University who was not involved in the study, because its level of brightness is just right for follow-up inspection across the electromagnetic spectrum. "It's bright enough for the instruments, but not too bright," he says. The James Webb Space Telescope, an infrared telescope set to launch next year, might be particularly suited to scope out KELT-9b. Gaudi is eager to see additional instruments take a crack at KELT-9b, such as the aging Hubble Space Telescope and NASA's Transiting Exoplanet Survey Satellite, which is also expected to launch in 2018. "This planet represents one of the best targets for characterizing any individual planet that we have to date," he says.

In Deming's mind, KELT-9b's accessibility is really what sets the planet apart from other wild finds. He laughs as he confesses he is glad KELT-9b is not habitable. "I'm interested in what I can measure," he explains. "Those habitable worlds are difficult to measure, and this will be measurable."

Gaudi and his team are intrigued by far more than just precise measurements, however. They have a whole list of KELT-9b attributes they would like to understand further. In fact, the planet is so strange, one of Gaudi's graduate students bet him a bottle of single-malt scotch that they had not even found a planet (clearly Gaudi won). KELT-9b is surprisingly puffy, given that its mass is almost three times Jupiter's: most ultraheavy worlds are several times more massive than Jupiter yet about the same diameter, compressed into Jupiter-sized orbs by their immense gravitational fields. KELT-9b, by contrast, is almost twice Jupiter's diameter. For reasons no one fully understands yet, the planet also orbits its star from pole to pole, rather than around the equator as would be expected. Gaudi assumes KELT-9b's atmosphere must be boiling off into space given how much heat it is receiving from the star, so he would like to know at what rate that is happening. One of Gaudi's colleagues, Thomas G. Beatty, an astronomer at the Pennsylvania State University, is most curious about what molecular mayhem is brewing on the dark side of the planet. The temperature drop outside the star's direct light is enough to tease hydrogen atoms into recombining, he says. The process is probably similar to the hottest types of welding on Earth. "That's where the real fun begins," Beatty says.

KELT-9b is also unique in its proximity to its giant, tempestuous A-type star. Previous research on older, cooler A-type stars found few large and tightly orbiting planets like KELT-9b—perhaps because as they cool and expand over time, the stars consume any unlucky closely held planets. Having found one very tightly orbiting planet around a young A-star, Gaudi thinks the hunt for other duos experiencing intermediary steps in star-planet evolution will help astronomers solve the mystery of why most older A-types lack companions.

Outer space does not give up its gems easily. So when something as wondrous as KELT-9b can also be thoroughly poked and prodded with today's technology, astronomers will struggle to resist its pull. While research groups submitted their project ideas for the James Webb Space Telescope months ago, Deming has a feeling a few teams will want to change course and pore over KELT-9b instead. If that is the case, the plethora of discoveries from the scorching planet are bound to expand what astronomers deemed possible. "Every time we lift up a rock, we find a new exoplanet, and we have to go back to the drawing board," says Gaudi. Until then, Gaudi has a few glasses of scotch to enjoy.

About the Author

Leslie Nemo is a freelance science journalist living in Brooklyn, N.Y. You can find more of her work at www.leslienemo.com.

Nearby Earth-Size Exoplanets May Have Water

By Charles Q. Choi

Roughly 39 light-years from Earth, the dim red star TRAPPIST-1 possesses a record-breaking seven Earth-size planets. Now astronomers using the Hubble Space Telescope have discovered hints that some of these planets could possess water, suggesting they may be suitable for life as we know it.

TRAPPIST-1, located in the constellation of Aquarius, is 2,000 times dimmer than our Sun, a bit less than half as warm, and only slightly larger in diameter than Jupiter. It is an "ultracool M dwarf," a kind of star that makes up about 15 percent of the stars neighboring our solar system.

Using TRAPPIST (TRAnsiting Planets and PlanetesImals Small Telescope), a telescope in Chile, in February 2017 astronomers discovered TRAPPIST-1 possessed the largest number of Earth-size worlds found so far in a single planetary system. At least three of these planets lie within the star's habitable zone, the so-called "Goldilocks zone" in which temperatures could be just right for liquid water to flow in rivers and pool in lakes, seas and oceans. Since there is life virtually wherever there is water on Earth, these findings raised the possibility that TRAPPIST-1's Earth-size planets could support life.

However, just because an Earth-size planet lies within a star's habitable zone does not mean it is actually habitable. A key factor in whether such worlds can support life is how much ultraviolet light they receive, since ultraviolet rays can destroy water and organic molecules.

Scientists used the Hubble Space Telescope to measure the amount of ultraviolet light emitted by TRAPPIST-1 and to estimate the amount of ultraviolet light that each planet in the system received. When they were first discovered, each of the seven worlds was assigned a letter in order of their discovery as a placeholder for other possible names, starting with "b" for the innermost and going to "h" for the outermost. Planets e, f and g all lie within TRAPPIST-1's putative habitable zone.

Lower-energy ultraviolet rays can break up water molecules into their atomic constituents, hydrogen and oxygen, while higher-energy ultraviolet light can heat up the upper atmosphere of a planet to allow that hydrogen and oxygen to escape into outer space. Moreover, planets especially close to their stars can suffer the same fate as Venus – intense starlight can turn all their water into steam, which is a potent heat-trapping greenhouse gas, leading to a steam-filled atmosphere and a runaway greenhouse effect that can bake a planet over hundreds of millions or billions of years until it is dry as a bone.

No one yet knows the exact age of TRAPPIST-1 and its planets–the current best estimate is anywhere between five and a half to nearly 10 billion years. Hubble's measurements of TRAPPIST-1's ultraviolet radiation suggest that, in that time, some of the star's planets could have lost vast amounts of water. For instance, planets TRAPPIST-1b to TRAPPIST-1d might now be in a runaway greenhouse phase, and planets within the orbit of TRAPPIST-1g may have each lost more than 20 Earth-ocean's worth of water.

Still, the researchers' calculations hint that not all hope is lost for the worlds of TRAPPIST-1: under certain conditions the star's outermost worlds, including e, f and g may have each lost less than 3 Earth-ocean's worth of water during their lifetimes. Consequently, they may still harbor substantial amounts of moisture.

"It's very promising that these outer planets could have retained their water," says Mark Marley, a planetary scientist at NASA Ames Research Center in Mountain View, California, who did not take part in the research. "They did not say that water could be there, but if you imagine the planets were water towers, they could say how much of a leak each had."

Moreover, when planets form, they can trap some of their water in their deep interiors, and "this water can then be released into the atmosphere over time through, for example, volcanism," says study lead author Vincent Bourrier, an astrophysicist at the Geneva Observatory in Switzerland. "We found that the outer planets would, as well, have been the ones likely to inject the most water into their atmospheres." The researchers detailed their findings online August 31 in *The Astronomical Journal.*

The researchers also looked for a giant hydrogen cloud surrounding TRAPPIST-1c that could have originated from water molecules in the planet's atmosphere being broken apart by ultraviolet light. Although they did not detect such a cloud, they plan to use Hubble again to search for such clouds around all of TRAPPIST-1's planets, Bourrier says.

As much as ultraviolet light could render some planets too dry for life around TRAPPIST-1 and other M dwarfs, another new study found a dearth of it may also make these red stars uninhabitable for life as we know it. Using computer models and the known properties of red dwarfs, researchers found that such stars might not emit enough of the long-wavelength, medium-energy ultraviolet light that likely helped trigger chemical processes key to the rise of biology on Earth.

"This illustrates the dangers of treating ultraviolet radiation as a monolith, when in fact different wavelengths of ultraviolet interact very differently with planets and putative life," says planetary scientist Sukrit Ranjan at the Harvard Smithsonian Center for Astrophysics in Cambridge, Massachusetts, the lead author of that study. He and his colleagues detailed these findings online July 10 in *The Astrophysical Journal*.

It remains uncertain whether the best hosts for habitable planets are ultracool M dwarfs such as TRAPPIST-1 or much warmer stars such as the sun, says Jeffrey Linsky, an astrophysicist at the University of Colorado at Boulder, who did not participate in either study. "The argument has gone back and forth like a pendulum several times and will likely continue to do so as we learn more about physical processes in planet atmospheres and the emission of host stars at all wavelengths," Linsky says. "It would not surprise me if the best host stars are intermediate in temperature."

About the Author

Charles Q. Choi is a frequent contributor to Scientific American. *His work has also appeared in* The New York Times, Science, Nature, Wired, *and* LiveScience, *among others. In his spare time, he has traveled to all seven continents.*

The Race to Find Alien Moons

By Rebecca Boyle

In 2007 a network of automated telescopes observed a star about 433 light-years away in the constellation Centaurus. The star dimmed noticeably for at least 54 days, with its feeblest light measured around April 29. In 2012 astronomers determined that this star hosted a huge, Saturn-size gaseous planet orbited by a spectacular array of 37 rings. Just as Saturn does, this world known as J1407b has a gap in its ring system. Scientists suggested the gap may indicate a moon with roughly the mass of Earth.

Until the end of the 20th century, the only known planets were the seven worlds with which Earth shares the sun. This situation changed with the earliest exoplanet discoveries in the late 1990s and was completely overturned starting in 2009, when the Kepler space telescope opened its lens. We now know that the cosmos is peppered with planets, that there are far more planets than stars, and that these worlds come in almost every imaginable size, location and type. Arguably for the first time since the days of Galileo Galilei, discoverer of Jupiter's largest moons, and astronomers such as Huygens himself, humans are seeing our place in the universe with fresh eyes. We have yet to find a distant planet that looks just like home or to confirm that an exoplanet is orbited by a moon of its own. But we are getting closer.

Astronomers began speculating about exomoons in the early 2000s, after several exoplanets began winking into distant starlight, and searches since 2018 have turned up a few promising candidates. Locating a moon outside our solar system would mark another reorientation of our cosmic perspective. We will learn whether moons are ubiquitous or rare; whether they are usually large or small, compared with their planets; whether they often form along with their planets or are created in later cataclysms; and whether they come in groups or typically fly solo. We will be able to understand whether our solar system is unique and whether Earth and its solitary, huge moon stand alone.

"Every time we see an exoplanet, I think it's a mirror on our own history," says Alex Teachey, a postdoctoral researcher at the Academia Sinica Institute of Astronomy and Astrophysics in Taiwan and co-discoverer of a possible exomoon of the exoplanet Kepler-1625b. "In what ways are we common and in what ways are we uncommon? Just as we started seeing with exotic exoplanet systems, we could be surprised in what we see with exomoons as well."

The Perks of a Moon

Earth remains unique, in this solar system and everywhere else we have looked so far. It is the only planet known to harbor life. It is the only planet whose active innards sculpt its outer face, in the form of plate tectonics, a process that itself plays a role in the dispersal and evolution of life. It is the only planet with an atmosphere thick enough to support liquid water, a climate that has remained stable for millennia and a just-right distance from its sun that keeps it warm but not too hot. These conditions exist at least in part because of Earth's moon.

The moon's role in Earth's history goes back to the very beginning, some 4.5 billion years ago, when a planet the size of present-day Mars collided with the infant Earth. The cataclysm left behind an incandescent, oblong Earth and a boiling moon. The moon has been cooling and moving away from Earth ever since. The planet became more spherical as the moon began to recede, and Earth's crust flexed under the resulting tidal force. The early crust deformed, possibly causing the onset of tectonics. The moon's recession also slows Earth's spin, lengthening our day by almost two milliseconds every century.

The moon's heft is significant relative to Earth's; our planet's mass is only 81 times greater than that of our satellite, a ratio many times smaller than for the other moons of our solar system. Saturn, for instance, outweighs Titan 4,200-fold. The moon's pull guards Earth's axis, keeping the planet near a constant 23.5-degree tilt with respect to the sun. This configuration protects Earth's climate over millennia, as opposed to Mars, which lacks sizable moons and thus wobbles on its axis between zero and 60 degrees at the extremes every few million years–a change that drives dramatic climatic shifts. The moon provides the primary influence over Earth's tides, which shape coastlines and the life in the oceans. Our moon's tides most likely played a role in evolution, shepherding the first plants and tetrapods from the salty marshes of the coasts and onto land.

The moon is more than a silent, spectral satellite; it is a world unto itself, which Earth's occupants have both used and contemplated since the first sighted beings looked skyward. Earth would not be Earth without the moon. Neither would the oceans, or poetry, or religion, or science, or any of us.

If Earth's singular past provides any prologue, exoplanets might be different without moons, too. Unmooned exoplanets might be lifeless rocks, doomed to tilt like Mars through the millennia, freezing or boiling and preventing atmospheres and life from holding on. Exomoons themselves, if they exist, might even be better places for life than their planets. The search for life beyond our solar system may need to focus on planets with the possibility of moons and even on the moons themselves.

Although Huygens could not have known it, his Titan, an orange ball of haze dotted with methane and ethane rivers and lakes, is quite planetlike. It would not be hospitable to us or any life we would recognize, but it contains liquid and an atmosphere, meaning it contains a chance for mixing of ingredients and a chance for life. Saturn, with its intense gravity and clouds of ammonia, would never be a safe haven for life; the same scenario exists for Jupiter, practically a half-star with absurd radiation belts and gas layers that would be fatal to us. But the same cannot be said of their moons.

"Because of our solar system, we know Jupiters can have significantly large moons that can have water," says Chris Fox, a graduate student at Western University in Ontario. "If you have a Jupiter in the habitable zone of its star, you could see a moon that is like Earth, and the moon may have life. Given the number of moons that there [likely] are, maybe life on moons is even more common than life on planets."

Exomoons, in other words, may be habitable themselves, and they might help make their host planets habitable, too. Finding them will bring us closer to understanding those worlds as well as our own.

An Intriguing Candidate

Astronomers suspected the cosmos was full of worlds and their moons long before the Kepler telescope saw its first light. In 1999 Paola Sartoretti and Jean Schneider, now both at the Paris Observatory, became the first to propose searching for exomoons using the transit method.

If a star and its planets are arrayed on a flat plane as viewed from Earth—as if you were looking at the solar system from the side rather than the top down—the star will appear to dim for a brief period because a planet has moved in front of it. This eclipselike frontal pass is called a transit. When transits repeat on a regular schedule, you can usually be confident an orbiting planet is causing the dimming. The Kepler telescope used this method to find planets for a decade. Sartoretti and Schneider argued that moons orbiting at wide distances from their host planets would be detectable in this way, too, if the moon were beside the planet at the time of the transit. A host star could dim more than usual or could even dim twice if the moon's orbit were sufficiently distant from the planet. If you were standing on the planet during the transit, for widely separated moons, the moon's phase would have to be near first or last quarter; a full moon or new moon, when the moon is directly opposed to the star, would not make a discernible dent in the planet's transit.

Even if exomoons orbited close to their host planets, making a secondary dimming implausible, astronomers might still detect moons by looking at how a planet's repeating transit pattern appears to change over time, Sartoretti and Schneider suggested. Often transits recur with metronomelike precision. But sometimes they are a little off, with a transit beginning or ending just a bit earlier or later than scientists predict—an effect called transit timing variation.

This can happen because other planets orbit the star and tug on one another, but it can also happen when a planet is hosting a large moon.

To understand why, it helps to understand that Earth's moon does not orbit our planet exactly. Rather both bodies orbit their mutual center of mass, called a barycenter. The barycenter is still located on Earth because the planet is more massive than the moon. (To be precise, the barycenter is located in Earth, in the mantle, offset from Earth's gravitational center.) As a result, Earth wobbles very slightly as it orbits the sun. This wobble is one thing Sartoretti and Schneider recommended looking for.

In 2017 Teachey and David Kipping, an astronomer at Columbia University, sifted through data from the Kepler telescope to look for any indications of a moon interfering with the star's light. They analyzed some 300 planets, hoping to find a population of moons. They found just one candidate: Kepler-1625b.

They applied for time on the Hubble Space Telescope and were surprised when they got it, both recalled. Then they studied the Hubble data for a year, with part of that time spent learning how to use them. When Kipping and Teachey finished their analysis, their Hubble observations showed that the planet's transit began sooner than it should have, implying a moon was by its side. The planet's transit time varied by about 20 minutes in five years of data. "We know there's something pushing that planet around," Kipping says, "and we think it's a moon."

Teachey and Kipping posted their paper to a preprint server in early 2018, and it was ultimately published in October 2018 in *Science Advances*. They say the evidence supports the existence of a Neptune-sized moon around Kepler-1625b, which itself is many times the size of Jupiter. Kipping and Teachey stopped short of claiming a discovery. "I think people were frustrated by the way we reported it," Teachey says. "People thought we were both trying to get credit for a discovery but also kind of covering our rear because we're not fully claiming it. I understand people's frustration–is it there or not there? But there's a lot of unknown unknowns."

Immediately following their initial announcement, other astronomers jumped into the fray. René Heller, an astronomer at the Max Planck Institute for Solar System Research in Göttingen, Germany, replicated part of Teachey's findings but found insufficient evidence for a moon. Laura Kreidberg, who studies exoplanet atmospheres, was unable to confirm a key part of the results. Kreidberg, now director of the new Atmospheric Physics of Exoplanets Department at the Max Planck Institute for Astronomy in Heidelberg, Germany, recalled having a friendly but somewhat awkward conversation with Teachey a few months after his paper published. "Alex worked so hard on this, and I don't want to minimize what he did," she recounted. "I sort of swooped in there with a bunch of years of experience using this instrument. I'm rooting for Alex. Both of us want the moon to be there. I mean, how cool would that be?"

The Race

Exomoon attention only grew in the months after Kipping and Teachey's announcement. Soon teams of researchers were poring over Kepler data on their own, trying to find transit variations that could indicate moons. Others turned to instruments such as the Very Large Telescope's Spectro-Polarimetric High-contrast Exoplanet REsearch tool (SPHERE). Cecilia Lazzoni, currently a postdoctoral fellow at the University of Exeter in England, claimed finding a giant exomoon using a SPHERE survey. In a paper published in *Astronomy & Astrophysics*, she described it as a companion to a very low-mass brown dwarf, a dim object between a planet and a star that does not fuse hydrogen but is many times the size of Jupiter. Lazzoni's world and its accompanying body may be more like binary giant planets than a world and a moon. If such objects are common, astronomers will have to grapple with how to define what constitutes a planet and a moon.

In 2019, Phil Sutton of the University of Lincoln in England reanalyzed the super-Saturn, J1407b. The planet and its rings were first discovered by Eric Mamajek, now at the NASA Jet Propulsion

Laboratory, and several colleagues. Sutton wanted to find evidence for moons that orbit outside the ring, as most of Saturn's do, so he set out to determine whether J1407b's 37 rings are sculpted in the same way. He could not find any evidence for external moons guiding the gaps in the rings and instead found that an external moon would likely shred the disk apart. "We all get excited that we might have actually found something," Sutton says, but reanalysis often dampens hopes. "It's just really tricky to confirm."

Then, in the summer of 2020, Fox, the Western University grad student, peered through more Kepler data. He and his adviser, Paul Wiegert, scrutinized 13 Kepler planets and found eight with transit timing variations that can be explained by exomoons. But, as Fox pointed out, the variations could be something else, too; the possibilities range from stellar activity such as flares to other planets. "In many cases, we were able to match the transit timing variation pattern with a moon, but in all the cases, we could explain them by the presence of a second planet," he says.

Teachey was critical of the work when Fox posted it to a preprint server that summer and equally critical of coverage that claimed Fox was describing moons when he really described moon-or-planet scenarios. In a young and rapidly growing field, especially one with high stakes for career-making discoveries, some growing pains are inevitable, Teachey says. "We're not trying to stifle people working on moons, and we don't want to come across like we're being gatekeepers," he says. "But at the same time, this is part of the game. We're going to call into question conclusions that in our view are not supported by the evidence." Fox's paper was published in 2021 in *Monthly Notices of the Royal Astronomical Society*.

In November 2020 Kipping hosted the first-ever exomoon meeting, an informal conference held over Zoom that brought together some 80 researchers from around the world. The scientists discussed new detection methods, theories for moon and exo-ring formation, new constraints on exomoon sizes and candidates, and related subjects. "I think we need to be a little more organized as a community to have a better shot," Kipping says.

A definitive detection remains elusive in part because the astronomers are all asking so much of their telescopes and their data. The wee blip in brightness that results from a planet transiting its star can be hard enough to see on its own. Comparing that with a shift in transit timing by just a few moments—remembering that the objects in question are hundreds of light-years away—is an excruciatingly exacting measurement.

Kreidberg says she is frustrated that she was unable to figure out why she and Teachey could not reach the same answer, using the same Hubble data. They shared each other's processing methods, and she tried to closely replicate his steps but could not reconcile the findings. "My only regret is that we weren't able to figure out what the difference was," she says. "The takeaway for me was, we are really pushing the limits of what Hubble can do. It was designed to look at faint distant galaxies, not nearby planets with moons. We're doing the best we can with data processing, but it's a fine art to pull the signal out."

Other challenges are geometric. Because of Kepler's laws (that is, Johannes Kepler, who discovered the rules governing planetary motion and for whom the planet-hunting telescope is named) and Newton's laws, moons' orbits are more stable within a certain distance from their planets, known as the Hill radius. The closer a planet orbits to its star, the likelier that the star's gravity will interrupt the moon's orbit, potentially sending it spiraling into the planet or out of the star system entirely. But the data from Kepler, Hubble and other observatories usually capture planets that orbit near their stars—often very near, closer than even Mercury is to the sun. Although these planets are relatively easier to find than planets farther away, they may be more likely to be moonless. "If we're looking at planets that are transiting their stars, we are looking at planets whose gravitational influence has been severely diminished and are less likely to host a moon," says Stephen Kane, a planetary astrophysicist at the University of California, Riverside. He published a paper in 2017 arguing that compact planetary systems, such as the TRAPPIST-1 system of seven terrestrial planets, are unlikely to host any moons at all.

Planets that wind up at greater distances from their stars, such as Jupiter and Saturn, are more likely to host moons, notes Alice Quillen, an astronomer at the University of Rochester, who has studied the super-Saturn J1407b. When a planet is more distant from the star, the star is less likely to fuss with the planet's gravity, allowing a moon to stay in place. If far-flung exoplanets resemble the solar system's own outer worlds in size and composition, they may also be likelier to grab planetary crumbs, errant asteroids and dwarf planets. Neptune's moon Triton is thought to be a captured dwarf planet from the distant Kuiper belt, a small world like Pluto, that was swept into Neptune's embrace after the solar system formed.

But large, distant planets are hard to find, partly because they take a long time to orbit their stars—one year on Jupiter, equaling one transit, takes almost 12 Earth years, meaning astronomers would have to watch such a planet for more than two decades before finding a definitive signal. And they are hard to spot because they can be confused with other objects. Periodic dims in a star's constant light are as likely to be caused by star pairs that occasionally move in front of each other as outer planets. "You can't find things in outer solar systems because it is too easy to confuse with eclipsing binaries," Quillen says. "You have to spend a lot of time trying to get rid of stuff that isn't what you want."

Stars themselves can also confuse the signals. The sun turns out to be a particularly quiescent star; other stars tend to be more active, churning out flares and radiation and developing spots that can also affect their apparent brightness. "The trouble with measuring the brightness of a star is that if you improve the precision too much further, you start to run into stellar activity," Kane says. "Stars will produce noise that is comparable or even greater than the signal expected from a moon. It essentially creates a ceiling you can't rise beyond, and that's a really big challenge."

Undaunted, some astronomers are turning to creative mathematical and observational methods. Apurva Oza, a lecturer at the University of Bern in Switzerland, is looking for an Io. The Jovian volcanic moon is visible with binoculars or a modest telescope; it is one of the four satellites Galileo Galilei observed in 1609. But viewed with sensitive instruments, Io is one of the most glaringly obvious objects in the sky. It positively emanates sodium and potassium, which it spews into space in vast quantities as Jupiter's gravity rends its innards and Io's volcanoes erupt. Io's exosphere can extend up to 500 times Jupiter's radius, Oza says. What is more, its signature would be visible no matter where the moon is located; an astronomer studying a transiting planet would not have to worry about the exomoon's phase. An exo-Io could be behind the planet, and its vast plasma cloud would still be detectable with the right instruments. "If you spray that gas everywhere, you just enshroud it, and you'll see it during transit," Oza says.

Spectrographs on several telescopes can already detect volatile gases within and around stars, he adds. Some have detected sodium, potassium and other signatures, which are often unexplained. "The missing factor could be a moon," Oza says. "It doesn't seem that outlandish when you think about it that way." He notes that an exo-Io would hardly be a place to look for any exobiology, however: "We're not looking for habitability here. We're looking for explosive environments, which is most of the universe."

The Way Forward

Astronomers hope the James Webb Space Telescope, which launched in December 2021, will be able to hunt for exomoons with greater precision. Kepler was designed to find Earth-sized planets around sunlike stars, so objects smaller than Earth can be difficult to pick out. Researchers frequently used the Spitzer Space Telescope, but it was retired in January 2020. Not much else on Earth or in the heavens can find exomoons at the moment, so astronomers are relying on better data-processing methods—and getting ready to wait.

"Part of our job remains not just looking for these things but coming up with better ways to look for them," Teachey says. "People think a discovery is a eureka moment. And it's more like, 'Let's see if it fails this test. And this test.' Then you're like, 'Well, it's kind of holding up.'"

Ground-based observatories such as the Extremely Large Telescope, under construction in Chile's Atacama Desert, could also spot exomoons under the right circumstances. The European space telescope PLAnetary Transits and Oscillations of stars (PLATO), set to launch in 2026, could help the search, too. Further into the future, satellites such as the Large UV/Optical/IR Surveyor (LUVOIR)–which might launch sometime in the mid-2030s–could provide excellent exomoon-hunting capabilities. But all these projects are still years away.

"With exomoons, it's kind of Hubble or bust right now," until James Webb gets going, Kipping says. In the meantime, he hopes the burgeoning exomoon community will continue figuring out new strategies for working with the data that exist so far. Kreidberg is also hopeful that James Webb will find exomoon signals but concedes that a definitive discovery may still be a while off.

"This is the cutting edge," Kreidberg says. "Figuring out what we know and how well we know it is an evolving process. You have to be an optimist to work on exoplanets." And, maybe, exomoons.

About the Author

Rebecca Boyle is a Scientific American *contributor and an award-winning freelance journalist in Colorado. Her new book,* Our Moon: How Earth's Celestial Companion Transformed the Planet, Guided Evolution, and Made Us Who We Are *(Random House), explores Earth's relation with its satellite.*

We Need to Widen the 'Habitable Zones' Seen around Alien Stars

By Phil Plait

Astronomers have, so far, discovered nearly 5,500 exoplanets–alien worlds orbiting alien stars–with more than 7,000 candidate planets still waiting to be confirmed.

That's a lot of planets.

We search for these distant worlds because we're curious and want to know what other planets are like. By understanding other worlds, we can better understand our own. But make no mistake, scientists are humans, and we want the answer to one of the biggest questions of the scientific age: Are we alone?

We may get an answer soon. Our technology is just on the cusp of detecting exoplanetary biosignatures, telltale signs of life such as molecules in planetary atmospheres that could indicate the presence of biology. Until then, though, it's useful to investigate everything we can about these planets and categorize them by their likelihood of supporting life.

A handy concept astronomers have for this is called the habitable zone. This is a region around a star where temperatures allow a planet to potentially harbor oceans, seas or lakes of liquid water on its surface. Too close and the stellar heat boils the water off. Too far and it freezes. Between these extremes, though, it could be just right. This middle-ground aspect is why some astronomers refer to it as the Goldilocks zone. (Although, to be honest, I've never cared for this moniker; I'd opt for the Baby Bear zone. Naming it after the person who broke into that ursine home and usurped all their hard-won food and property smacks of colonialism.)

As an idea, it's pretty useful. All life on Earth needs liquid water, and because we're unaware as yet of any other way life might occur, that's a good place to start.

Measuring habitable zones, however, isn't straightforward. Calculating the radiation a planet receives from its star is easy; that depends on well-understood physics. The hard part is the planet itself. A dark planet absorbs more light and heats up, whereas a lighter-toned one will reflect more light and be colder.

A planet's atmosphere plays an even bigger role: if it's loaded with greenhouse gases, then the planet needs to be farther out from the star to be clement. Just look at Venus–our sibling world in many ways, similar to Earth in size and mass–to see how important that is. Lead would melt on our evil twin's surface because of its overwhelmingly thick atmospheric blanket of carbon dioxide. It's not exactly an Earth-like planet.

So on its own, finding a planet in a star's nominal habitable zone is no guarantee it will be, well, *habitable*, even if it's a small (and presumably) rocky world like our own. A lot more must be known, including whether it even has an atmosphere, what that atmosphere is composed of, and more. This problem is so complicated that astronomers argue over where our sun's habitable zone even starts, and we're literally inside it. What's more, the habitable zone might not be the only place where liquid water can exist in a solar system.

In the 1970s the *Voyager 2* spacecraft flew by Jupiter's moon Europa and saw surface features that hinted at the presence of a liquid-water ocean underneath its frozen surface. We've since collected extremely compelling evidence of subsurface Europan liquid water kept warm by the moon's interaction with Jupiter's immense gravity.

Europa isn't the only ocean moon, either. In 2005 images from the Cassini spacecraft showed huge plumes of water erupting off the surface of Saturn's icy moon Enceladus. Probably generated by Saturn's tidal activity, similar to what warms Europa, these geysers point to the presence of immense pockets of subsurface liquid water, if not another ocean.

Now we think there could be many such sub-ice oceans inside the moons of the solar system's outer planets and even in some of the larger objects orbiting the sun past Neptune.

In fact, we know of quite a few rogue planets meandering through interstellar space, which were probably ejected from their original planetary systems as the worlds there first formed and gravitationally interacted with one another. The ones we find tend to be gas giants even more massive than Jupiter. If they have icy moons, those, too, could be heated enough to have subsurface oceans. So *you might not even need to have a star* to have a habitable world!

To throw even more cold water on habitable zones, there are other liquids to ponder as well. Saturn's largest moon Titan is too cold to have liquid water on its surface, but Cassini observations in 2006 showed vast lakes of liquid methane there. Methane is a carbon-based molecule, so many of the ingredients for life are perforce there. Who knows whether there are alien fishies swimming in Titanic lakes a billion kilometers from the sun?

Needless to say, these are well outside the sun's quote-unquote habitable zone, yet it's possible that life may abound inside these prima facie frozen moons. Clearly, the concept of a habitable zone is woefully incomplete for the purpose of determining where life might exist. So is it time to kick it to the galactic curb?

Let's not throw out the extraterrestrial baby with the subsurface bathwater! Some years ago a team of planetary astronomers wrote that the term needs modifying and suggested it be replaced with "temperate zone." I think this idea is a fine one, still highly useful if we're looking for Earth-like planets, which, to be clear, *we are*. It might not include frozen moons of gas-giant planets or worlds where life as we *don't* know it may evolve, but as long as we are aware of these limitations, it's still handy. Renaming the concept could solve some things.

The habitable zone, even by any other name, was never meant to be an ironclad rule. It was always a guideline, an idea to inform astronomers that they might be onto something interesting when a planet is found in a certain spot. It's not a device for measuring potentially habitable worlds so much as a way to bookmark them for future observations.

It's easy to want to draw lines in the sand–a planet *this* far from its star is inhospitable, but a planet *that* far is great–but nature almost never behaves that way. It generally works on a spectrum, with fuzzy borders and even larger overlaps. It's always good to keep that in mind when reading about scientific discoveries.

A planet being in its star's habitable zone might not be sufficient, or even necessary, to make it habitable, but it's still a pretty good place to start when looking for life. We just need to make sure we don't stop there.

About the Author

Phil Plait is a professional astronomer and science communicator in Virginia. He writes the Bad Astronomy Newsletter.

Section 3: The Milky Way and Beyond

The First Milky Way Black Hole Image Lets Scientists Test Physics

By Seth Fletcher

Deep in the heart of the Milky Way, strange things happen. This is a place where stars slingshot around apparently empty space at an appreciable fraction of the speed of light. Scientists have long thought that only a supermassive black hole could explain the stars' movements, but until this year, they hesitated to say that outright. For example, when astronomers Reinhard Genzel and Andrea Ghez shared a portion of the 2020 Nobel Prize in Physics, their citation specified that they were awarded for "the discovery of a supermassive compact object at the centre of our galaxy," not the revelation of a "black hole." The object is known as Sagittarius A* ("Sagittarius A star").

This past spring, however, the astronomers behind the Event Horizon Telescope (EHT) settled the matter by unveiling the first image of a supermassive black hole at the center of the Milky Way. It wasn't the first picture of a black hole this collaboration had captured–that was the iconic image of M87*, which they revealed in April 2019. But it was the one they wanted most. Sagittarius A* is our own private supermassive black hole, the still point around which our galaxy revolves.

Black holes trap everything that falls in, including light, so they are, in a very real sense, unseeable. But they warp spacetime around them so severely that, when they are illuminated by glowing streams of infalling matter shredded in their gravitational grip, they cast a "shadow." The shadow is about two and a half times larger than a black hole's event horizon, the boundary in spacetime through which nothing that passes can ever return.

The EHT captured images of this shadow using a technique called very long baseline interferometry (VLBI), which combines radio observatories on multiple continents to form a virtual Earth-size telescope, an instrument with the highest resolution in all of astronomy. In April 2017 the EHT collaboration spent several nights pointing that virtual instrument at Sagittarius A* and other supermassive black holes. The scientists then spent years analyzing the raw data and converting them into an image.

Part of the reason it took so long was the global devastation of the COVID pandemic. But the bigger challenge was that Sagittarius A* is constantly changing. The observatory's previous target, M87*, the black hole at the heart of the galaxy Messier 87 (M87), is so huge that the matter swirling around it takes many hours to complete a full orbit. Practically speaking, that means you can stare at it for a long time, and it will scarcely change. Sagittarius A* is more than 1,000 times less massive, so its appearance changes about 1,000 times faster, as matter moves in tighter, quicker orbits around the black hole. Katie Bouman, a California Institute of Technology computer scientist and astronomer who co-leads the EHT's imaging working group, said that matter orbits Sagittarius A* so quickly that it changes "minute to minute." Imagine taking a time-lapse photograph of a speeding bullet–it's not easy.

If Sagittarius A*'s mercurial nature made it hard to see, it also makes it an ideal laboratory for understanding black holes and Einstein's general theory of relativity, his hallowed theory of gravity. Through decades of study with all manner of telescopes, astronomers already knew Sagittarius A*'s basic measurements (its mass, diameter and distance from Earth) to great accuracy. Now, at last, they've gained the ability to watch it evolve–to watch as it feeds on flaring, flashing streams of matter–in real time.

Scientists started to suspect that a black hole lurked in the heart of the Milky Way in the early 1960s, not long after the discovery of active galactic nuclei–extremely bright regions at the cores of some faraway galaxies illuminated by voraciously feeding supermassive black holes. From our perspective here on Earth, active galactic nuclei are a thing of the past–we see them only in the distant, ancient universe. Where did they all go? In 1969 English astrophysicist Donald Lynden-Bell argued that they didn't go anywhere. Instead, he said, they just went to sleep. Dormant supermassive black holes, he predicted, are slumbering all around us in the hearts of nearly all spiral galaxies, including our own.

In 1974 American astronomers Bruce Balick and Robert Brown pointed radio telescopes in Green Bank, W. Va., at the center of the Milky Way and discovered a dim speck they suspected was our galaxy's central black hole. They found it in a slice of sky, known as Sagittarius A, within the constellation Sagittarius. Radiation from the new source was lighting up—or "exciting"—surrounding clouds of hydrogen. Brown borrowed from the nomenclature of atomic physics, in which excited atoms are marked with an asterisk, and named the newfound speck Sagittarius A*.

For the next two decades radio astronomers kept gradually improving their view of Sagittarius A*, but they were limited by a lack of suitable telescopes, relatively clunky technology (think reel-to-reel magnetic tape) and the inherent difficulty of looking into the crowded galactic center.

Sagittarius A* is concealed by a multilayered veil. The first layer is the galactic plane—27,000 light-years' worth of stars, gas and dust that blocks visible light. Radio waves sail through the galactic plane unimpeded, but they're obscured by the veil's second layer—the scattering screen, a turbulent patch of space where density variations in the interstellar medium knock radio waves slightly off course. The final layer concealing Sagittarius A* is the infalling matter surrounding the black hole itself. Peering through that barrier is a bit like peeling off an onion's skins. The outer layers of matter emit longer-wavelength light—the same wavelengths VLBI traditionally works with. Making VLBI work with shorter-wavelength light would enable closer-in views approaching the black hole's event horizon, but it was a major technological challenge.

For a while, astronomers using other techniques besides VLBI had more success, steadily gathering indirect evidence that Sagittarius A*'s "speck" was actually a seething supermassive black hole. In the 1980s physicist Charles Townes and his colleagues showed that gas clouds in the galactic center were moving in ways that made sense only if they were under the influence of some great, unseen gravitating mass. And in the 1990s Ghez and Genzel independently began tracking the orbits of giant blue stars in the galactic center, mapping their motion around a heavy but hidden pivot point.

Meanwhile the situation for radio astronomers improved. In the late 1990s and early 2000s a new generation of short-wavelength radio telescopes started to come online–telescopes that, if augmented with lots of bespoke equipment, could participate in VLBI observations at the microwave frequencies thought to shine from the edge of Sagittarius A*'s shadow. At the same time, the computing revolution that led to solid-state hard drives and smartphones in every pocket vastly increased the amount of data that each observatory in a network of radio telescopes could record and process.

In 2007 a small precursor to the EHT took advantage of these trends and used a trio of telescopes in Hawaii, California and Arizona to pierce the veil surrounding Sagittarius A*. The result was far from an image, but the project saw something–presumably, light from the long-sought shadow.

The first prediction of black hole shadows came in 1973, when physicist James Bardeen showed that a black hole in front of a bright background would produce a silhouette. He decided that "there seems to be no hope of observing this effect." In 2000, however, astrophysicists Heino Falcke, Fulvio Melia and Eric Agol demonstrated that a microwave-gathering, Earth-size radio telescope should be able to see the shadow of Sagittarius A*.

Half a decade afterward, a few dozen of the astronomers and astrophysicists laboring in this obscure corner of astronomy agreed on the formal goal of building a virtual planet-scale radio telescope to observe that shadow. The first official kickoff meeting for the project occurred in January 2012, and the Event Horizon Telescope was born.

Five years later, after growing into a collaboration of more than 200 scientists with eight participating observatories across the globe, the EHT took its first realistic shot at seeing the shadow of Sagittarius A*. Over the course of 10 days in April 2017, telescopes in North America, South America, Hawaii, Europe and Antarctica collectively zoomed in on the galactic center and other black holes, gathering 65 hours of data on 1,024 eight-terabyte hard drives, which were shipped to supercomputer banks in Massachusetts and Germany for correlation.

Five years after *that*, the elated EHT researchers showed the world that their experiment worked.

The day the EHT collaboration revealed Sagittarius A*'s image, the *Astrophysical Journal Letters* published a special issue devoted to the new results. In six technical papers, the scientists presented a multidimensional portrait of our black hole.

The EHT image confirmed the basics. We've long known that Sagittarius A* is about 27,000 light-years away. Years spent tracking the orbits of stars around Sagittarius A* with infrared telescopes had already given astronomers an accurate measure of the black hole's mass—the equivalent of roughly four million suns. Plug those two numbers (distance and mass) into equations derived from general relativity, and you can calculate the expected size of the black hole's shadow. Sure enough, the image matches the prediction. The shadow has a diameter of 52 micro arc seconds, which means that to us here on Earth it is, in the astronomers' formulation, the size of a "doughnut on the moon." After seeing the shadows of both Sagittarius A* and M87*—black holes that differ in mass by three orders of magnitude—the scientists concluded that the phenomena are "universal features of black holes."

The EHT observations, combined with simultaneous monitoring by the Chandra and Nu-STAR x-ray telescopes and other instruments, are starting to settle long-standing questions about Sagittarius A*'s environment. By measuring the spectra of light shining from the object—that is, the light broken up into its constituent frequencies—astronomers long ago determined that the matter orbiting the black hole is a diffuse gas of electrons and protons. We now have a much better idea of where that matter comes from. Observations from the Chandra x-ray telescope show that the black hole pulls that matter from the atmospheres of stars orbiting it. Not that it pulls very much. Sagittarius A* is on a starvation diet—less than 1 percent of the stuff captured by the black hole's gravity ever makes it to the event horizon. That explains why the black hole is so dim. Despite being four million times more massive than our sun, Sagittarius A* is only 100 times brighter.

It wasn't always so faint. As recently as 60 or 70 years ago, Sagittarius A* appears to have gone on a feeding binge, a burst of activity that left "light echoes"–light bouncing off nearby clouds of dust and gas–that astronomers have detected using x-ray telescopes. And it still has its active moments. On April 11, 2017, Sagittarius A* emitted a bright x-ray flare. The cause, most likely, was the twisting and recombining of magnetic fields inside the matter orbiting the black hole–the same basic dynamic that causes solar flares in our own star. Directly observing these flares is a major goal for future campaigns.

There's much more to learn about Sagittarius A*. For example, the EHT observations show that, like everything else in space, the black hole is spinning–but we don't know how fast. Future observations also aim to show exactly how the black hole eats infalling matter, and, with luck, make movies of the black hole evolving over time.

The first EHT picture of Sagittarius A* is just the beginning, but it does tell us what the object is not. The presence of a distinctive shadow means Sagittarius A* has an event horizon–the defining feature of a black hole. That means we finally know it's not a really, really, really dense star, or a wormhole, or a naked singularity (a point of infinite density unconcealed by an event horizon), or any of the other exotic oddities theorists have proposed over the years. It's nothing stranger than a supermassive black hole–still a pretty strange thing indeed–now brought within view, closer to revealing its secrets.

About the Author

Seth Fletcher is chief features editor at Scientific American. *His book* Einstein's Shadow *(Ecco, 2018), on the Event Horizon Telescope and the quest to take the first picture of a black hole, was excerpted in the* New York Times Magazine *and named a* New York Times Book Review *Editor's Choice. His book* Bottled Lightning *(2011) was the first definitive account of the invention of the lithium-ion battery and the 21st century rebirth of the electric car. His writing has appeared in the* New York Times Magazine, *the* New York Times *op-ed page,* Popular Science, Fortune, Men's Journal, Outside *and other publications. His television and radio appearances have included CBS's Face the Nation, NPR's Fresh Air, the BBC World Service, and NPR's Morning Edition, Science Friday, Marketplace and The Takeaway. He has a master's degree from the Missouri School of Journalism and bachelor's degrees in English and philosophy from the University of Missouri.*

Where (Some of) Earth's Gold Came From

By Imre Bartos and Szabolcs Márka

The first detection of a neutron-star collision took 60 years after Margaret Burbidge and colleagues realized that the heaviest elements in the universe must be created in violent cosmic explosions. Gravitational waves observed from the collision by LIGO and Virgo in 2017 led to a worldwide hunt for the cosmic flash produced by the neutron star debris. The findings elevated neutron stars to be leading candidates to create Burbidge's elements.

Some of the elements from Burbidge's process are radioactive; they steadily decay after the initial explosion that created them. These elements act as omnipresent cosmic clocks that carry information about their origin. We found that such clocks, present in the solar system, point to a neutron star collision that occurred a mere 1,000 light-years from what is today the solar system, a distance that is 1 percent the size of the Milky Way. This single event is responsible for the creation of 0.3 percent of some of the heaviest elements found on Earth today, including gold, uranium and iodine.

Matter deposited by this single event is ubiquitous. Every human has an eyelash worth of matter in them from this collision, mostly in the form of iodine. A gold wedding ring contains 10 milligrams. A Tesla Model 3 has five grams. A nuclear reactor has 200 kilograms. Of the gold ever refined by humanity, 600 tons are from this single collision.

Neutron stars are ultracompact dead stars, formed when massive stars collapse under their own gravitational pull. They weigh as much as our sun, but their size is that of New York City. They are in effect a gigantic atomic nucleus, made up primarily of neutrons, hence their name.

Scientists long suspected that neutron stars occasionally collide with each other in the universe, even though definite proof was elusive. Freeman Dyson suggested in the early 1960s that two neutron stars may be closely orbiting each other; he was interested in the extraction of energy from such a system in humanity's distant future. Russell Hulse and Joe Taylor were the first to detect two close-by neutron stars, in the 1970s. These neutron stars are set to collide in 300 million years; their distance is continuously decreasing as a result of the emission of gravitational waves.

The first hint of neutron-star collisions came as a serendipitous side effect of the Cold War. The United States launched a network of military satellites, called Vela, to monitor the Soviet Union's compliance with a nuclear test ban treaty during the 1960s. Instead of rogue detonations on Earth, the satellites detected mysterious signals from outer space. Some of these cosmic flashes of energetic photons, called gamma rays, later turned out to be telltale signs of colliding neutron stars billions of light-years away.

In 2017, the LIGO detectors in the U.S. and the Virgo detector in Europe observed a gravitational wave signal that was identified as a neutron star collision 130 million light-years from Earth, proving that such collisions do happen in the universe. This was a distant event, but the fact that LIGO and Virgo detected it after monitoring the sky for only about four months meant that neutron stars regularly collide. In our Milky Way galaxy, there is one collision every 100,000 years. If we had millions of years to wait, we would eventually encounter a collision by chance not too far from our solar system.

When and where neutron stars collide has particular importance for the birth of the solar system. Before the sun and the planets were formed, matter that eventually made them was in the form of a primordial cloud made of gas and dust. This cloud was many times greater than the solar system today, casting a giant cosmic net that collected atoms traveling between stars in the so-called interstellar medium. Some of the matter accumulated in this presolar cloud came from neutron star collisions, including most of the gold, platinum, uranium and other heavy elements we find on Earth today.

To reconstruct what happened, we can turn to radioactive isotopes that were produced by neutron stars and were deposited in the presolar cloud. While many of these isotopes have long since decayed, their presence in the early solar system is recorded in meteorites that were formed soon after the presolar cloud collapsed and began to form the sun and the planets.

The information encoded in radioactive isotopes is revealing. We found that the relative quantities of several short-lived isotopes with half-lives less than 100 million years is typical if they originate from neutron stars, but they are inconsistent with another kind of cosmic explosion: a supernova, produced by massive dying stars. Supernova explosions are 1,000 times more frequent in the Milky Way than neutron star collisions.

This means that the shortest-lived isotopes would be regularly replenished if produced by supernovae, making them certain to be present at the time of the solar system's formation. For the rarer neutron star collisions, the shortest-lived isotopes are depleted soon after a merger, and stay depleted until the next. This means that it is probable that at the time of the solar system's formation, this isotope will be depleted. The observed abundances of the short-lived curium-247 and iodine-129 isotopes in the early solar system show this depletion, ruling out supernovae.

At the time of Burbidge's original work and in the decades that followed, supernovae were considered to be the main candidates for sources of the heaviest elements. This paradigm has been challenged by multiple lines of evidence, including the optical signal observed from the neutron star collision detected by LIGO and Virgo, and our work.

Radioactive isotopes also point to a collision near the presolar cloud. Comparing the measured quantities to numerical simulations of the Milky Way, we found that a single collision likely produced a substantial fraction of the short-lived elements of the early solar system. For example, curium, whose longest-lived isotope has a half-life of 15.6 million years, was mainly produced by a single collision.

Elements that come from neutron stars play important roles in our lives. They include valuable metals, such as gold and platinum, or elements essential to making modern electronic devices. Some of them, such as iodine, are critical to life itself. Had the rate of collisions been different, or had Earth been at a different part of the Milky Way, the abundance of neutron-star matter on Earth, and with it our environment, may have been very different.

With the advent of new types of observatories, such as those capable of detecting gravitational waves or neutrinos, our exploration of the universe increasingly relies on combining information from different cosmic messengers. In this new field called multimessenger astrophysics, radioactive elements found in the solar system are a valuable addition. There is much to learn about the most extreme cosmic events and how they shaped our environment here on Earth.

This article was published in Scientific American*'s former blog network and reflects the views of the author, not necessarily those of* Scientific American

About the Authors

Imre Bartos is an associate professor in the Institute for High Energy Physics and Astrophysics at the University of Florida and a member of the LIGO Scientific Collaboration, the LISA Consortium and the Cosmic Explorer Consortium.

Szabolcs Márka is the Walter O. LeCroy, Jr. Professor of Physics at Columbia University and a member of the LIGO Scientific Collaboration.

Mystery Object Blurs Line between Neutron Stars and Black Holes

By Charlie Wood

For more than a decade astrophysicists have wondered why nature appears to show an odd restraint in the way it slays stars. In life, they range from pip-squeaks to behemoths. Small ones simply burn out and fade away, but something more curious happens to the jumbo-size variety. When such a star dies, its great bulk causes its innards to implode as a core-collapse supernova. The process sparks a cataclysmic explosion and compresses some of the remains into astrophysical exotica—often a neutron star or, for the very heaviest suns, a black hole. Yet a pronounced rift appears to divide the weight classes of these two types of massive stellar corpses. Although astronomers have spotted neutron stars weighing up to around two solar masses and black holes as light as five, middleweight cadavers have gone entirely missing—until now.

In June 2020 the Laser Interferometer Gravitational-wave Observatory (LIGO) Scientific Collaboration announced the first conclusive detection of a stellar remnant falling into the so-called mass gap between neutron stars and black holes. After months of calculations, researchers at LIGO and the Virgo gravitational-wave detector in Italy concluded that such waves rippling through Earth in August 2019—an event dubbed GW190814 that was initially classified as a black hole consuming a neutron star—actually came from a 23-solar-mass black hole swallowing a mysterious 2.6-solar-mass object. Whether the smaller body is the heaviest known neutron star or the lightest known black hole—or a truly exotic beast, such as a star made of particles distinct from those of normal stars—its existence suggests that the theories describing the most extreme stellar fates need updating.

"I would rank this as definitely the most exciting announcement we've seen from LIGO since the original binary black hole discovery and then the first detection of a neutron star collision," says Duncan Brown, a gravitational-wave astronomer at Syracuse University, who was not involved in the research. "We're probing a new piece of astrophysical understanding of the universe."

The new finding hints that the cosmos may enjoy a wider freedom in how it disposes of stars than researchers had supposed. Whether that leeway means atomic building blocks have enough brawn to support more monstrous neutron stars or that supernovae can forge tinier black holes, LIGO's detection shrinks the gulf between those two most plausible scenarios.

"The idea of a mass gap as a true gap with nothing in it, I think, is getting progressively destroyed," says Philippe Landry, a LIGO member at California State University, Fullerton. "This is going to be one nail in the coffin."

From a fundamental physics perspective, the line separating neutron stars from black holes is razor-thin. If you toss an apple onto a neutron star at the limit of what its constituent neutrons can bear, it will abruptly collapse into a black hole. The heftiest known neutron star weighs 2.14 times the mass of our sun. And nuclear theorists suspect the objects can grow somewhat heavier, with the most optimistic models putting the complete breakdown of matter at 2.5 solar masses. Based on such theories, the LIGO collaboration calculated that the chance of the lighter partner in GW190814 being a neutron star is less than 3 percent. A neutron star that heavy, Landry says, would be a "complete game changer."

Although they would never be able to collect any winnings, most astrophysicists would bet that 2019's merger involved a big black hole gobbling up an improbably tiny one. But while nuclear theory makes that scenario more plausible than one involving a single black hole and a neutron star, it still challenges the best theories of how such systems come to be. "Basically," Landry says, "something's got to give."

As far as astrophysicists know, making a pint-size black hole–and then feeding it to a larger one–should pose a nearly insurmountable obstacle to the universe. A conceivable way of creating such a diminutive partner is to mash together two bulky neutron stars, an event LIGO witnessed in 2017. What are the chances, however, that a special matchmaking environment, such as a chaotic galactic center dense with stellar corpses, brought two neutron stars together–and also managed to set up the resulting mass-gap black hole with a much larger companion? "There's nothing that forbids it," says Feryal Özel, an astrophysicist at the University of Arizona, who is not involved with LIGO. "But it's more of a soap opera."

To many, the simplest option is that the mini black hole was born–as most black holes are–directly from the heart of a dying star. The basics of stellar death are simple: a star blows off a massive shell, leaving its core to crumple into a black hole or neutron star. But predicting the exact aftermath of a messy explosion involving gravity, thermodynamics and particle physics represents something of a cosmic final exam that astrophysicists are still working on. "Pick up the comprehensive encyclopedia of physics," Brown says, "and you will probably need almost every piece of physics in there to model a supernova."

So when a team led by Özel analyzed groups of known neutron stars and black holes in 2010 and concluded that none were likely to lie between two and five solar masses, supernova researchers jumped at the chance to place more meaningful limits on the inscrutable process. A rough picture emerged in which fast, violent explosions expelled most of a star's material cleanly, leaving the naked core to contract into a classic neutron star of perhaps two solar masses. In gentler cataclysms, however, some debris failed to escape and crashed back onto the neutron star. This material might (conveniently) add at least three solar masses, producing a black hole weighing–at minimum–as much as five suns.

Thus, the vacant zone became a tool for theorists, says Vicky Kalogera, an astrophysicist at Northwestern University and a LIGO member. She and her colleagues asked themselves, "What do I need to do to the core-collapse mechanism to create the gap?" she says.

LIGO's mass-gap discovery hints that perhaps they need not bother: a 2.6-solar-mass stellar black hole would suggest that no such rigid regulations apply. Supernovae may be free to search out the fundamental line between neutron stars and black holes after all, removing one onerous stipulation from the astrophysical final exam. "I think this thing pretty much says there isn't a mass gap," Brown says.

Nevertheless, the divide between neutron stars and black holes may well persist as a cosmic proclivity rather than a rule. If you looked naively at star sizes, Özel says, you would conclude that black holes should be everywhere. Their relative absence still suggests to her that supernovae likely conspire against them to some degree. "It could be that these objects are very difficult to make, but once in a blue moon, a supernova explosion lands you there," she says. Plus, researchers will still have to figure out why a mass gap seems to separate black hole and neutron star duos seen in x-rays but not those detected by gravitational waves.

Hard answers will come only with more detections of seemingly impossible objects. LIGO can tell tiny black holes from big neutron stars if the partners have a similar mass or give off a visible flash when they merge (colliding black holes are not expected to detonate with a burst of light, as merging neutron stars do, at least in principle). Even if ambiguous events continue to pile up, however, just watching where the masses fall will reveal a lot about what nature does with its leftover stars.

"After one weird system is discovered, then we need more of them," Kalogera says.

About the Author

Charlie Wood is a journalist covering discoveries in the physical sciences both on and off the planet.

Brown Dwarfs Could Reveal Secrets of Planet and Star Formation

By Katelyn Allers

Breathe. Breathe. I repeated these words to myself like a mantra. At 18,400 feet, my body was craving oxygen, and I had to concentrate on pulling enough air into my lungs. I was on the summit of Cerro Toco, a stratovolcano overlooking Chile's Chajnantor Plateau, now home to the Atacama Large Millimeter/submillimeter Array, one of the world's premier radio telescopes. Between the thin atmosphere and the barren red terrain of the mountain, it felt like I was on Mars. My colleagues and I were testing the atmospheric conditions on Cerro Toco. If they were good enough, they might justify taking on the technical challenges of building an observatory at such a remote, high-altitude site.

Earth's atmosphere is a problem for astronomers, and clouds frustrate many an observer. Atmospheric turbulence smears starlight, making stars appear to dance and flicker when close to the horizon. Molecules such as water vapor and carbon dioxide in the atmosphere absorb incoming starlight, particularly infrared light. With more than half of Earth's air below the summit of Cerro Toco (a point repeatedly raised by my burning lungs), we hoped that new and exciting insights could come from a dedicated infrared telescope there.

The sense of adventure that had led me to this summit had also sparked my fascination with infrared astronomy, where scientists peer at the cosmos in light too red for the human eye to see. Infrared light tends to come from the dimmest and most distant objects observable. One class of objects best seen in the infrared is brown dwarfs. When I was in graduate school in the early 2000s, these bodies had only recently been discovered, and they presented many tempting mysteries. I came to be captivated by these uncanny orbs, which, in terms of their classification, occupy a boundary zone between stars and planets. I wondered where and how they formed and what they were like. I learned through my research that in addition to being interesting in their own right, brown dwarfs serve as an important bridge to our understanding of both planets and stars, with temperatures and masses intermediate between the two. Now I and other brown dwarf astronomers are enjoying a sweet spot for research–there are still many brown dwarfs waiting to be discovered, and we can build on the wealth of previous research to uncover new details of physical processes at work on these objects. We finally have the technological tools to study the atmospheres of brown dwarfs, for example, as well as their wind and rotation speeds, and to try to determine whether they might even host planets of their own.

In-Between Objects

Most stars are powered by the fusion of hydrogen into helium, a wonderfully stable process that keeps stars burning at the same temperature and brightness for billions of years. But if a would-be star never reaches high-enough temperatures or pressures to sustain hydrogen fusion, it is a brown dwarf, with a maximum mass of 8 percent of our sun's, or about 80 times the mass of Jupiter.

Recent studies indicate that brown dwarfs are nearly as common as stars, and they are everywhere. Brown dwarfs have been found in stellar nurseries alongside young protostars. They have been found in binary systems paired with white dwarfs, having survived potential engulfment by the white dwarf's previous red giant form. (Our sun, a yellow dwarf star, will one day turn into a bloated red giant, and after it dies, it will become a white dwarf.) Some of the closest stellar systems to our sun are brown dwarfs–the third and fourth nearest extrasolar systems, at 6.5 and 7.3 light-years, respectively (the closest are Alpha Centauri and Barnard's star). And yet, despite their ubiquity, most people have never heard of brown dwarfs.

Although they lack hydrogen fusion, brown dwarfs do emit light–thermal radiation from the heat within them. They start out relatively hot (around 5,000 degrees Fahrenheit), and over the subsequent billions of years, they cool and dim. Brown dwarfs never die; they spend eternity cooling off and fading away. The coldest known brown dwarf checks in at a temperature below the freezing point of water. Because they are so cool, most of the light they emit is at infrared wavelengths. They are far too faint for the unaided human eye to see in our night sky, but if we could look at them up close, they would probably have a dull orange-red or magenta hue.

In the more than two decades since astronomers began studying brown dwarfs, we have formed a fairly clear picture of their basic characteristics. Like our sun, brown dwarfs are composed almost entirely of hydrogen. The temperatures in their upper atmospheres are cool enough, however, that a variety of molecules can form. Signatures of water vapor are seen in nearly all brown dwarfs. As they cool further, their atmospheric chemistry changes, and different molecules and clouds become predominant. The evolution of a brown dwarf's atmosphere depends on its mass and age. Imagine a brown dwarf with a mass 40 times that of Jupiter, for instance. For the first 100 million years, it will have an atmospheric composition similar to that of a red dwarf star, with titanium oxide and carbon monoxide present in the mix. Between 100 million and 500 million years, the atmosphere will cool, and dusty clouds made of minerals such as enstatite and quartz will form. Roughly a billion years after that, the clouds will break up and sink, and methane will become the dominant molecular species in the upper atmosphere. The coolest known brown dwarf shows evidence of water-ice clouds, as well as water vapor and methane. We expect its atmosphere to contain significant amounts of ammonia, similar to what we see on Jupiter.

Beyond these properties, however, there are many things about brown dwarfs that we do not yet know. The mysterious nature of these objects has inspired some far-fetched ideas. Brown dwarfs were once considered to be a possible reservoir of dark matter, although this idea was quickly abandoned when it became clear that brown dwarfs emit light (that is, they are not dark) and that their contribution to the total mass of our galaxy is small. More recently, scientists proposed that life could form in the cool upper regions of brown dwarfs' atmospheres—an idea that brown dwarf experts quickly squashed because the dynamics are such that any life-form would cycle into deeper layers of the atmosphere that are hot and inhospitable.

And then there was the hoax of the Nibiru cataclysm, a prophesy put forward in 1995 that predicted an imminent, disastrous encounter between Earth and a brown dwarf. Astronomers would be very excited to see a brown dwarf up close, but there is no scientific evidence to support this doomsday scenario, and a brown dwarf would be visible for hundreds or thousands of years prior to any close encounter.

The First Brown Dwarfs

Scientists predicted brown dwarfs in the 1960s based on what they knew about how stars and planets form. It seemed that this intermediate category should exist, but astronomers were not finding any such objects in the sky. It turned out that brown dwarfs are simply very, very faint, and most of the light they emit is infrared. And infrared technology was still in its infancy–just not up to the task. Then came the year 1995, a big one for astronomy. Astronomers Michel Mayor and Didier Queloz found 51 Pegasi b, the first exoplanet known to be orbiting a regular star. Perhaps more important, at least to this highly biased author, the first brown dwarfs were discovered.

Teide 1 was identified in the famous Pleiades star cluster. Astronomers Rafael Rebolo López, María Rosa Zapatero-Osorio and Eduardo L. Martín first spotted it in optical images from the 0.80-meter telescope at the Teide Observatory in the Canary Islands. The object was young, still glowing slightly from its formation. The team observed the signatures of several elements in its atmosphere, including lithium. Stars usually burn up lithium as soon as they form, so this amazing detection proved that nuclear fusion was not occurring. They published their finding in September 1995.

Two months later astronomers announced the discovery of a second brown dwarf, Gliese 229B, a companion to another star. A group of astronomers at the California Institute of Technology and Johns Hopkins University first saw the object in an infrared image from the Palomar Observatory. They immediately knew that it was strange. It had unusual colors and displayed the signature of methane in its atmosphere. Conditions must be very cold for methane to be present because the highly reactive molecule usually turns into carbon monoxide at higher temperatures. Later observations revealed that the brown dwarf is about the same width as Jupiter, with a diameter of nearly 129,000 kilometers, but much denser, with 70 times as much mass.

By the time I started graduate school in 2000, we knew of more brown dwarfs, though not that many. I was focused on building infrared instruments, and I needed a subject for my research topic. My Ph.D. adviser studied star formation, so I decided to search for brown dwarfs in star-forming regions. I ended up discovering a good number of brown dwarfs in my thesis work, including some that were the first known to have masses putting them near the range of planets. At the time we had no idea how these things formed, and we did not know whether there was a lower-mass threshold, but we started finding smaller and smaller objects.

All in all, my thesis work published fewer than 20 new brown dwarf discoveries, but they made a significant contribution to the total number known. Since then, new instruments have found many, many more. The main contributors were the 2 Micron All Sky Survey (2MASS), an infrared survey conducted in the early 2000s, and the Wide-field Infrared Survey Explorer (WISE), a space telescope launched in 2009. The current tally of brown dwarfs is about 3,000. There are many more to be found, though–estimates suggest that the Milky Way contains between 25 billion and 100 billion brown dwarfs.

Formation Scenarios

As the lowest-mass outcome of the star-formation process, brown dwarfs offer astronomers a unique chance to deepen our understanding of the basic steps involved in the birth of stars and planets. Stars form in complexes of gas (mostly molecular hydrogen) and dust known as molecular clouds. If a molecular cloud contains enough mass, gravity can overcome the gas pressure supporting the cloud and cause it to collapse into a star. During the collapse, any small amount of rotation in the cloud becomes amplified, much like how ice skaters spin faster when they pull their arms in. This rotation of the cloud material leads to the formation of a circumstellar disk of matter surrounding the nascent star, which then becomes a crucible for planet formation.

When brown dwarfs were first discovered, astronomers assumed they might form in a process similar to that for stars, but they were perplexed as to how the gravity from such a small mass was able to overcome gas pressure and initiate a collapse. In writing this article, I looked back over some grant and telescope proposals from early in my career, most of which were aimed at better understanding the formation mechanism of brown dwarfs. At the time there were several competing ideas. Some theories involved disrupting the formation of a star before it had reached its final mass. Perhaps some process physically removed the brown dwarf or burned off its natal environment, leaving behind a miniature star?

Other hypotheses invoked a scaled-down version of star formation or a scaled-up version of planet formation. This is a lovely example of using a variety of possible theories to make distinct, testable predictions. As we discovered the ubiquity of circumstellar disks around brown dwarfs, determined the distribution of stellar and brown dwarf masses in a variety of environments, and mapped the orbits of brown dwarfs in binary pairs, it became clear that most brown dwarfs seem to form like scaled-down stars–but from a smaller reservoir of gas. And the fact that brown dwarfs form circumstellar disks raises the tantalizing possibility that they host planets. Although we have never seen any for sure, it is very likely that planets grow in these disks just as they do around stars. Scientists hope the coming years will finally see the confirmed discovery of worlds orbiting brown dwarfs.

Recently researchers discovered isolated brown dwarfs with masses similar to those of giant planets (less than 13 times the mass of Jupiter), which again raised the question of how they might have formed. Could some of these planetary-mass brown dwarfs have arisen in the circumstellar disks of more massive stars–in other words, formed just as planets do?

To test the mechanism for the formation of planetlike masses, my colleagues and I proposed a survey with the Hubble Space Telescope. Because Hubble is in orbit, it avoids the smearing and absorption of light by Earth's atmosphere, which makes it ideal for imaging binary pairs of brown dwarfs. Through this survey, in 2020 we discovered a unique system of brown dwarfs that strongly supports a starlike-formation mechanism for planetlike masses. The system, Oph 98 AB, is very young in cosmic terms (three million years old), and its two components weigh in at 15 and eight times the mass of Jupiter. These extremely low-mass objects are separated by 200 times the distance between Earth and the sun. Because Oph 98 A and B are so light and so widely separated, the system has the lowest gravitational binding energy of any known binary pair. The weak binding energy means that these bodies must have formed in their current orientation, rather than originating elsewhere and later becoming a pair, which points to a starlike-formation mechanism. And the young age of the system (yes, we consider three million years young!) means that planetary-mass objects apparently do not take any longer to form than stars.

New Insights

Brown dwarf science has now reached a stage where we are able to make more precise measurements and ask more detailed questions than ever before about these still mysterious objects. Among the most interesting recent discoveries are the coldest brown dwarfs, known as Y dwarfs. These objects have temperatures ranging from 350 degrees F down to –10 degrees F. I love to joke when working on Y dwarfs that I am studying the coolest systems in the galaxy! Though not quite as cold as Jupiter (–234 degrees F), these Y dwarfs have enabled us to make the first meaningful comparison between brown dwarfs and the atmospheres of the giant planets in our solar system. Y dwarfs are difficult to observe because they are both cool and very dim. The light they do emit is predominantly in the infrared range, at wavelengths of three to five microns, where Earth's atmosphere makes observations difficult.

Regardless, my colleagues and I have published spectra of several Y dwarfs and used theoretical models to infer the presence of water-ice clouds, as well as a significant amount of vertical mixing in the atmosphere. In this same wavelength range, Jupiter emits its own light (rather than just reflecting the light of our sun) and shows significant vertical mixing as well. Our hope is that by studying Y dwarfs, we will be able to disentangle properties of Jupiter that come from its planetary nature—in other words, the fact that it formed in the circumstellar disk of our sun and is constantly illuminated by sunlight—and properties that may be ubiquitous among cool gaseous objects, be they planets, exoplanets or brown dwarfs. Thus far our studies are showing that highly dynamic atmospheres tend to be the norm.

These insights about brown dwarf atmospheres have led to a new subfield: exometeorology. Although brown dwarfs are too far away for us to visually examine their atmospheric features, we can see their imprint through changes in brightness. As a cloud or other feature rotates in and out of view, it changes the light coming from the brown dwarf. Astronomers have analyzed the brightness variations of brown dwarfs over many rotations and have created maps of their spots and bands, which look remarkably like the familiar stripes and storms on the giant planets in our own solar system. Some brown dwarfs have been found to change in brightness by up to 25 percent over one rotation. The results of these studies are leading us to better understand atmospheric processes more generally—we have found that brown dwarfs with temperatures at which clouds break up show large variations in brightness and that young objects tend to show greater variability in brightness.

Scientists have also discovered other similarities between brown dwarfs and gas giants. Both, for example, tend to have strong magnetic fields and aurorae, as revealed by radio observations of the signatures of charged particles spiraling in their magnetic fields. The measured magnetic field strengths for brown dwarfs are 1,000 times stronger than Jupiter's magnetic field and 10,000 times stronger than Earth's. I like to imagine what the night sky might look like from one of these brown dwarfs–given the beauty of Earth's aurora borealis, it would likely be a spectacular sight.

Recently a student's question prompted another project to examine how the atmospheres on brown dwarfs compare with those on planets. When I teach courses in introductory astronomy, we cover the planets of the solar system (and of course, I sprinkle in a lot of information about brown dwarfs as well). A tidbit I present is that the length of a Jovian day depends on how you measure it. If you clock the motion of visible features in Jupiter's equatorial region, you measure a rotation period that is five minutes shorter than the rotation period measured in the radio signal, which probes its interior rotation. A student asked me why this difference in rotation period occurs, and I replied that it was because Jupiter's equatorial features are pushed along by strong zonal winds. The winds on Earth are driven by the redistribution of solar energy, but we are not sure to what degree this applies to Jupiter's winds.

After the lecture, I started thinking about this further. Astronomers have measured radio emission in brown dwarfs, which occurs via the same mechanism as Jupiter's radio emission, so we can measure an interior rotation period. And we can use our method of monitoring brightness changes to measure the atmosphere's rotation period. Thus, I hatched an idea to measure the wind speed on a brown dwarf for the first time. The best candidate we had to try out the technique was a methane brown dwarf with confirmed radio emission. To determine the wind speed, we would need to measure both periods to a precision of less than 30 seconds. My colleagues and I submitted a proposal to use the Spitzer Space Telescope to measure the brown dwarf's brightness variations and applied to use the Karl G. Jansky Very Large Array in New Mexico to measure a more precise radio period. It still feels like a small miracle that our measurements revealed a period difference of just more than a minute, which equates to a wind speed of 2,300 kilometers per hour. We published our findings last year in the journal *Science*. This high wind speed on an isolated brown dwarf means that atmospheric winds are not always driven by the redistribution of solar energy, leaving open the question of whether Jupiter's winds are driven by the sun.

Astronomers continue to search for more brown dwarfs. Some surveys focus on identifying large samples of brown dwarfs via deep imaging surveys of the whole sky such as 2MASS, WISE, and the Panoramic Survey Telescope and Rapid Response System (Pan-STARRS). Citizen scientists have also become involved in the search through projects such as Backyard Worlds, which allows anyone to examine WISE data for signs of brown dwarfs and other moving objects. We expect that upcoming large surveys with the Vera C. Rubin Observatory (due to begin observing early next year) and the Nancy Grace Roman Space Telescope (launching in 2025) will work to further complete our census of brown dwarfs.

Sadly, we could not get funding for the telescope on Cerro Toco, and it was never built. But once the James Webb Space Telescope is launched later this year, astronomers will have an unprecedented look at brown dwarfs in the infrared, without interference from Earth's atmosphere. The first cycle of observations planned includes programs to study the atmospheric chemistry of Y dwarfs and the cloud composition of dusty brown dwarfs and even a search for planetary systems around brown dwarfs. Exciting times are certainly ahead for those of us who study some of the cosmos's most overlooked objects.

About the Author

Katelyn Allers is an astronomer whose research focuses on low-mass stars and brown dwarfs. Until recently, she was a professor at Bucknell University. She now produces digital-training materials at Northwest Registered Agent.

Mysterious Fast Radio Bursts Are Finally Coming into Focus

By Adam Mann

Nobody noticed when an Australian radio telescope captured a fleeting explosion coming from far beyond the Milky Way in 2001. Records of the powerful flare sat unseen for more than half a decade until a group of scientists sifting through archival data spotted the eruption—a so-called fast radio burst (FRB). According to one of those scientists, astrophysicist Duncan Lorimer of West Virginia University, the burst produced as much energy in a few thousandths of a second as the sun does in a month.

Today researchers know that these explosions happen at least 800 times a day all over the sky, and they are one of the most active topics in astrophysics. Although much about FRBs remains unknown, in just the past year a clearer picture has started to emerge. "I think we're closer to understanding what some FRBs are," says Ziggy Pleunis, an astrophysicist at the University of Toronto. "But as we've been going on this quest, new discoveries have led to new questions."

The study of FRBs is now at an inflection point. A torrent of new detections and deeper studies have elevated certain models of their inner workings while eliminating others, and several upcoming projects should help further narrow the possibilities. Meanwhile scientists have learned that the bright light of an FRB carries within it a record of the contents of the intergalactic depths it traversed along its way to Earth, providing information about galaxies and the material between them that no other mechanism can.

Magnetic Moments

In April 2020 three separate research teams detected an enormous blast of radio energy coming from a magnetar located in the Milky Way. Magnetars are an extreme kind of neutron star, a city-sized remnant left behind when a massive star dies in a supernova. A magnetar's magnetic field can be so strong that approaching within 1,000 kilometers of one would disrupt your body's atomic nuclei and electrons, causing you to effectively dissolve.

Magnetars were already a leading candidate for the source of FRBs. But the few dozen known magnetars in our galaxy had never been observed to produce eruptions that might resemble the phenomena. The discovery of a short and formidable radio burst from a galactic magnetar called SGR 1935+2154 was exactly what researchers had been missing. If the same burst had come from another galaxy, its signature would have been indistinguishable from a typical FRB. "That was a huge moment for the field," says Kenzie Nimmo, an astronomer at the University of Amsterdam. "It alleviated all doubt that at least some FRBs come from magnetars."

How, exactly, a magnetar can produce an FRB is still the subject of debate. Most theories involve some kind of jarring starquake or an explosion caused when a magnetar's twisting magnetic field lines snap and reconnect. Events such as these could directly generate an FRB's flash, or they might make a shock wave that heats up surrounding material, incinerating dust and turning gas into plasma to produce light as it travels outward.

Several telescopes saw an x-ray flash arriving just after SGR 1935+2154's radio signal, suggesting that whatever released the radio energy also generated more complicated side effects. Many details are still unclear. "Did this happen on the surface of the star, or in the magnetosphere, or in the material around the magnetar?" asks Emily Petroff, an astrophysicist also at the University of Amsterdam. "We still don't really agree on that."

Cosmic Curiosities

Because FRBs can vary in brightness, duration and other properties, it is unlikely that any single observation can explain them all. In the summer of 2021 the Canadian Hydrogen Intensity Mapping Experiment (CHIME), a dedicated FRB-hunting telescope in British

Columbia, released a catalog of 536 FRBs that it had detected during the first year of its operation, quadrupling the number known. The bursts were already known to come in two distinct flavors—those that repeatedly flash their signals and those that are one-off events. CHIME's data showed that nonrepeaters were far more common than repeaters and that each had different characteristics.

On average, the bursts from repeaters lasted longer than their one-off counterparts and emitted their light in a narrower range of frequencies. Whether the disparity represents a different mechanism for these flashes or something else about their progenitors' ages or environments remains to be seen. The situation resembles an earlier mystery surrounding another class of cosmic explosions: gamma-ray bursts, which were shown in the 1990s to arise from three separate types of events. Scientists hope to discover whether FRBs also fall into distinct populations with their own origin stories.

CHIME's catalog includes large numbers of FRBs from a variety of galaxies, muddying the link to magnetars, which emerge almost exclusively in galaxies that are churning out massive, short-lived stars. CHIME's FRB haul, however, includes many sources from quieter galaxies that are barely forming any new stars at all. "Magnetars can explain some fraction of FRBs. Nobody would dispute that," says Shami Chatterjee, an astronomer at Cornell University. "But is that all of them? Almost certainly not."

A paper published in *Nature* in February 2022 adds support to this assertion. Using an array of radio telescopes called the European Very Long Baseline Interferometry (VLBI) Network, a team determined the position of a repeating burst designated FRB 20200120E with extreme precision. The object had originally been localized to the nearby spiral galaxy M81, but VLBI zoomed in farther and revealed that it lives within an ancient hive of densely packed stars known as a globular cluster. Such collections mainly host stars around 10 billion years old—yet magnetars are thought to endure for only 10,000 years or so before lapsing into more sedate neutron stars. "This is a game changer," says Mohammadtaher Safarzadeh, a theoretical astrophysicist at Harvard University.

"Whatever is causing the FRB signal likely has the same age as the globular cluster and is definitely not a magnetar."

Magnetars might occasionally arise from two neutron stars crashing into each other—a production mechanism that has never been confirmed—which could potentially explain one in a globular cluster, says theoretical astrophysicist Bing Zhang of the University of Nevada, Las Vegas. But nobody knows exactly how often such events occur or how long the resulting magnetars would remain active.

Further complicating the magnetar picture is another curiosity: FRB 20180916B, also known as R3 because it was the third repeating FRB ever discovered. Originally pinpointed to a region toward the star-forming center of a spiral galaxy around half a billion light-years away, R3 was subsequently shown to be in the galaxy's outskirts, suggesting that it is either an older object or one somehow kicked away from its birthplace. Even stranger, this burst produces explosions only during a four- to five-day window of activity that occurs every 16.35 days, making it a so-called periodic repeater.

Researchers have been scratching their heads over its peculiar regularity. A magnetar that spins around on its axis like a top, sometimes pointing its blasts toward Earth and other times facing away, is one possible explanation. Another is a bursting object orbiting a second structure, such as a black hole surrounded by a disk of material, that cyclically obscures the explosive events. Scientists have even suggested it comes from a pair of orbiting neutron stars whose magnetospheres periodically interact, creating a cavity where eruptions can take place. "What makes the field so fun right now is that there are so many exciting possibilities," Chatterjee says.

Approaching Answers

FRB astronomers are still pursuing major questions. Are nonrepeaters really one-time events, or could they burst again if we watch for long enough? The magnetar in our galaxy appears to be fairly quiet. But was it significantly more active in its younger years? Could other esoteric scenarios, such as asteroids hitting a black hole, somehow produce FRB-like signals? Scientists are publishing new observations and theories all the time.

The CHIME collaboration is building a set of smaller telescopes that will help triangulate the exact positions of many FRBs. In a few years researchers expect to know the precise locations of hundreds or even 1,000 events. In addition to elucidating FRBs, these data will allow scientists to perform important measurements of the universe.

Astronomers first knew FRBs were coming from outside the Milky Way because their light was dispersed, meaning the higher frequencies arrived a few milliseconds before the lower ones. This pattern offers information about the matter the radio waves traveled through as they made their way through space. Astronomers believe there is much more regular matter in the universe than what we see in stars and galaxies, and they suspect that the missing matter lies in the intergalactic medium. In 2020 a team studied a handful of FRBs to estimate how much material their light passed through and showed it was almost exactly equivalent to the amount of matter expected there.

The ultimate goal is to use FRBs to map the matter throughout the universe. And light from some FRBs is highly polarized–its waves have been rotated by magnetic fields during its flight–potentially revealing information about magnetic conditions in other galaxies or the spaces between them. In the meantime, the mystery of FRBs' origins remains. "I fully anticipate, within the next decade, we'll get one or two more surprises, like the galactic magnetar that we didn't even know we should be looking for, which will push our understanding forward in a massive way," Petroff says.

If some nonrepeating FRBs arise from cataclysmic events such as neutron stars crashing together, as many astronomers suspect, they would also create gravitational waves. Were a radio telescope to see a blast at the same time as the Laser Interferometer Gravitational-wave Observatory (LIGO) or its counterparts around the world, it would sway some toward that possibility. And if such a collision produced a magnetar, could the initial cataclysmic one-off FRB give rise to a distinct, repeating FRB source? Time will tell.

Given recent history, more FRB excitement is likely in the coming years, Lorimer says: "Just when you think things are settling down, you have a year with all these remarkable discoveries."

About the Author

Adam Mann is a journalist specializing in astronomy and physics. His work has appeared in National Geographic, *the* Wall Street Journal, Wired, *and elsewhere.*

Section 4: Cosmology Explains the Universe

Time Flowed Five Times Slower Shortly after the Big Bang

By Lucy Tu

In a first, astronomers have observed time ticking in slow motion in the early cosmos, confirming Albert Einstein's century-old ideas about the reality-warping effects of our universe's expansion.

By tracking the flickering glows of luminous matter swirling through galaxies when the universe was a mere one billion years old (less than a tenth of its present age), two researchers have found that events back then appear to have unfolded at a pace five times slower than normal. Their findings were published earlier this month in *Nature Astronomy*.

"For decades Isaac Newton gave us this vision of a universe where space and time is fixed, and every clock across the universe ticks at exactly the same rate. Then Einstein shattered this vision by proposing that time is actually rubbery and relative," says Geraint Lewis, an astrophysicist at the University of Sydney and lead author of the study. "Now we've shown that Einstein was, once again, correct."

The Einsteinian concept of time running slower in the early universe arose in the late 1920s as astronomers were discovering cosmic expansion. Galaxies in the sky were found to be flying away from the Milky Way at high speed, swept along by the ceaselessly growing void–and the farther off they were, the faster they flew. This not only meant that the universe was once much smaller and denser–arising in a "big bang" from some compact, primordial point–but also that the most distant galaxies visible to us should be receding at close to the speed of light.

According to Einstein's special and general theories of relativity, both circumstances alter the flow of time. As light from one of those far-distant galaxies travels from the heavier gravitational grip of the deep, dense early cosmos and across the continuously expanding universe, it must traverse increasingly greater expanses of space to reach Earth. Consequently, time becomes stretched in a phenomenon known as time dilation: a clock running 10 billion years ago would tick at a normal rate to an observer from that time, but from the perspective of someone today, it would appear to be ticking much slower.

Astronomers had previously validated this slow-motion cosmos about halfway back through the universe's 13.8-billion-year history by examining the light from massive exploding stars called supernovae that detonated six to seven billion years ago. But such supernovae are too faint to observe at the immense distances needed to probe earlier cosmic epochs.

So Lewis and astrostatistician Brendon Brewer instead investigated much larger, more luminous objects known as quasars—bright astrophysical beacons formed from supermassive black holes gorging on gas at the centers of distant galaxies. Gas piles up and spirals around as it funnels at nearly light speed into a feeding black hole, where it heats up to temperatures of several trillion degrees Fahrenheit and emits an incandescent glow visible across the cosmos.

But this glow isn't steady. Black holes are messy, fitful eaters—and trillion-degree gas can go down less like a smooth milkshake and more like chunky peanut butter. Although this variability makes quasars easier to identify, it complicates their use as standard markers of cosmic time. If supernovae are akin to a firework, burning bright and quickly fading away, then quasars change brightness more like the stock market, with an unpredictable pattern of turbulent flickers. In fact, prior studies have failed to find a time dilation effect between quasars very distant from us and ones relatively close by.

"Those early findings inspired some fringe cosmologists to question whether quasars' variability adheres to our existing models of the universe. There were even suggestions that our long-held, fundamental idea that the universe is expanding was wrong," Lewis says. He adds that these studies used small samples or observed quasars over a short time period.

In contrast, Lewis and Brewer used a new, much more expansive data set: they looked at 190 quasars in all, covering a range of cosmic time from about 2.5 billion to 12 billion years ago. Each quasar's flickering was observed hundreds of times at multiple wavelengths across a span of two decades.

The duo also grouped the quasars by intrinsic luminosity. "We boxed bright quasars with bright quasars and faint quasars with faint quasars," Lewis says. This approach minimized the chance of making "apples-to-oranges" comparisons between distinctly different quasar types and allowed the researchers to calibrate each quasar's "ticks," yielding more certainty that some of the observed discrepancies in light fluctuations were caused by time dilation.

Ultimately, the researchers found that the tick-tock of the quasar clocks behaved just as Einstein's relativity predicts. Quasars found in faraway galaxies ticked slower than ones born in the later, nearby universe, with time dilation making those most distant appear to run at a glacial one fifth of the standard speed.

Katie Mack, an astrophysicist who holds the Hawking Chair in Cosmology and Science Communication at the Perimeter Institute for Theoretical Physics in Ontario, says that these findings provide clarity on various uncertainties surrounding quasar behavior. Specifically, the study confirms that quasars align with consensus expectations–and it reinforces the need for astronomers to consider time dilation when studying them.

"This is the first time that the effect of time dilation has been clearly observed with quasars, and it's comforting to know that there's nothing bizarre happening there," says Mack, who was not affiliated with the study.

While astronomers had anticipated the presence of the effect in the ancient universe, this prediction still needed to be tested. Michael Hawkins, an emeritus researcher at the University of Edinburgh's Institute for Astronomy, says the study serves as a valuable reminder for scientists to avoid complacency with established cosmological models, adding that Einstein's theory of general relativity upended centuries of science when it was introduced. Hawkins himself has previously conducted research that failed to detect time dilation in quasars, which he says underscores the significance of ongoing investigation and refinement in the field.

"To uphold scientific practice, you have to maintain skepticism until the very end, so it's critical to continue testing even the most well-established theories of the universe," Hawkins says. As a next step, he would like to see future studies replicate the analysis with a larger sample of quasars originating from galaxies even deeper in the cosmic past.

For Lewis, the work is more than a vindication of Einstein and modern cosmology. Accurate timestamping of ancient quasars might also prove useful for further exploring the nature of dark energy, the mysterious force thought responsible for a surprising acceleration in the universe's expansion.

"Standardizing and confirming our models is ultimately a step into the next generation," Lewis says. "The goal now is to chart the expansion of the universe in as much detail as possible."

About the Author

Lucy Tu is a freelance writer and a Rhodes Scholar studying reproductive medicine and law. She was a 2023 AAAS Mass Media Fellow at Scientific American.

Are Telescopes the Only Way to Find Dark Matter?

By Chanda Prescod-Weinstein

How do you think the dark matter problem is solved?" Vera C. Rubin urgently asked me, within minutes of being introduced at a 2009 Women in Astronomy conference. To this day, I can't remember what I said in response. I was awestruck: the famed astronomer who had won the National Medal of Science for her work finding the first conclusive evidence for dark matter's existence was asking me, a twentysomething Ph.D. student, for my opinion. I am certain that whatever I came up with was not very good because it was a problem that I had, until that moment, given no serious thought to. Until Rubin asked me my opinion, it had never occurred to me that I was entitled to have an opinion on the question at all.

If I disappointed her with my answer, she didn't show it. Instead she asked me to sit down to lunch with her and some other women astronomers, including former NASA administrator Nancy Grace Roman. Rubin then proceeded to fangirl over Roman, who is often referred to as "the mother of the Hubble Space Telescope." It was quite a moment for me, to watch an elderly woman who had uncovered one of the greatest scientific mysteries of our time excitedly introduce us to her own hero.

Rubin cemented her legacy in the 1960s, when she studied stars inside galaxies and found something odd: stars on the outskirts of galaxies were moving faster than they were supposed to, as if there was an invisible matter there contributing a gravitational pull. Her work echoed findings from galaxy cluster studies in the early 1930s by Fritz Zwicky, which had led him to suggest the existence of *Dunkle Materie*, German for "dark matter." Throughout the 1970s Rubin and astronomer Kent Ford published data consistent with this conclusion, and by the early 1980s scientists were in widespread agreement that physics had a dark matter problem.

Most attempts to track down dark matter in the laboratory have fallen into three categories. So-called direct detection experiments look for evidence of dark matter particles interacting with particles of normal matter–for instance, the element xenon–through one of the nongravitational fundamental forces, the weak force, as well as through hypothesized new forces. Collider experiments, such as those at the Large Hadron Collider near Geneva, take the opposite approach, smashing two regular particles together with the hope of producing dark matter particles. Meanwhile "indirect detection" experiments look for evidence of dark matter interacting with itself, with the resulting collision producing observable particles.

So far none of these strategies has turned up the missing matter. We still don't know if dark matter can talk to regular matter in any way beyond gravity. It may be impossible to produce in the accelerators we can build or to detect in the experiments we can construct. For this reason, astronomical observations–cosmic probes of dark matter–are one of our best hopes. These probes allow us to look for signatures of dark matter in environments that are difficult for us to produce on Earth–for example, inside neutron stars. More broadly, such searches look at dark matter's behavior under gravity in a variety of locations.

Despite the promise of this approach for studying dark matter, it has sometimes been caught in the middle between the astronomy and physics communities. Physicists tend to emphasize colliders and laboratory experiments and don't always prioritize links to astrophysical work. Astronomers tend to write dark matter off as a particle physics problem. This disconnect has implications for funding. In 2022 we have an opportunity to change that. The start of the 2020s marked the beginning of an important process known as the Snowmass Particle Physics Community Planning Exercise. This project, which takes place about once a decade, brings physicists together to explain prospective scientific projects to a congressionally mandated panel that will determine scientific priorities. For the first time, cosmic probes of dark matter will be a distinct topic of consideration. Although Snowmass does not make formal policy recommendations, it is certainly the case that at each stage of the organizational hierarchy, there will be decisions about what science to emphasize.

A Universe of Dark Matter Candidates

There is still much we don't know about dark matter, but we have come a long way since Rubin's work in the 1970s and 1980s. We now have good evidence to suggest that every galaxy lives in its own bubble of dark matter–called a dark matter halo–that extends well beyond the visible part of the galaxy. The amount of dark matter in these galaxy-halo systems surpasses the amount of matter in the stars and planets and gas. In other words, all the particles that we have been able to identify in labs and colliders–referred to collectively as the Standard Model of particle physics–contribute only about 20 percent of the normally gravitating matter in the universe. If we take into account dark energy and the fact that matter and energy are fundamentally equivalent, we are down to understanding only about 4 percent of the cosmos. The Standard Model is both a stunning achievement and a theory that is, apparently, deeply incomplete. We need a new particle or particles to solve the problem.

Physicists now have an assortment of dark matter candidates. Most scientists favor candidates that are "cold dark matter"–particles that move relatively slowly (meaning, at nonrelativistic speeds much slower than that of light). Within the class of cold dark matter, one of the classic models is the weakly interacting massive particle (WIMP). Scientists presume that WIMPs would have formed naturally in the early universe and predict that they have some kind of interaction with regular matter through the weak force. The most popular WIMP models fall into a category of particles called fermions–a class that includes electrons and quarks.

WIMPs were the most highly favored dark matter candidates for a long time, particularly in the U.S. Opinions have shifted in recent years, though, as evidence for WIMPs has failed to show up at the Large Hadron Collider or in any of the direct and indirect detection experiments.

Recently the particle physics community has become excited about another hypothetical dark matter candidate: an axion. Axions are predicted to have smaller masses than WIMPs, and they are not fermions. Instead axions belong to a class of particles called bosons—the category that includes photons, or particles of light. As bosons, axions have fundamentally different properties than WIMPs, which opens the door to an intriguing possibility about the structures they could form. Axions are what first drew me into the world of dark matter research.

Alluring Alternatives to WIMPs

Five years passed between my conversation with Vera Rubin and my first attempt at answering the question she had put to me. By then it was 2014, and I was a Dr. Martin Luther King, Jr., postdoctoral fellow at the Massachusetts Institute of Technology, appointed first to the Kavli Institute for Astrophysics and Space Research and then the Center for Theoretical Physics (CTP) and looking for something interesting to work on. It was there that Mark Hertzberg—at the time also a postdoctoral researcher at CTP—and I first started talking about a debate that had erupted among physicists: Could axions form an exotic state known from atomic physics called a Bose-Einstein condensate?

This possibility arises from a fundamental difference between bosons and fermions. Fermions must obey the Pauli exclusion principle, which means two fermions cannot share the same quantum state. This rule is why electron orbitals in chemistry can be so complicated: because the electrons orbiting an atom cannot occupy the same quantum state, they must spread out in different patterns with different amounts of energy called orbitals. Axions, on the other hand, can share a quantum state. This means that when we cool them enough they can all enter the same low-energy state and act collectively like one superparticle—a Bose-Einstein condensate. The possibility that this could happen naturally in space is, in my view, quite exciting.

Axions had been proposed in the 1970s by Hertzberg's Ph.D. adviser at M.I.T., Frank Wilczek, one of the first to realize that one consequence of a model proposed by Helen Quinn and the late Roberto Peccei was a particle, which Wilczek named "axion" after a brand of laundry detergent. Thus, Hertzberg was already quite familiar with axions. I, on the other hand, was relatively new to this idea. I had spent most of my career focused on other questions, and I had to get up to speed. Along the way, I learned to distinguish between the traditional axion and the class of particles that physicists have come to loosely refer to as axionlike particles.

The traditional axion arises from the Peccei-Quinn extension to the theory of quantum chromodynamics (QCD), which describes another of the four fundamental forces, the strong force. Although QCD is a highly successful model, it also predicts phenomena we've never observed. Peccei and Quinn's work solves this problem, while providing a mechanism for producing dark matter. But another idea called string theory also proposes a series of particles with the same mathematical structure as the original axion; these particles have come to be called axionlike. The traditional QCD axion is usually expected to have a mass of about 10^{-35} kilogram–several orders of magnitude lighter than the electron–but the larger class of axions from string theory can be much lighter, down to 10–63 kilogram.

The work Hertzberg and I did together with our postdoctoral adviser Alan Guth led us to quibble with a popular view of how axions might form Bose-Einstein condensates. A distinguished physicist, Pierre Sikivie of the University of Florida, had prompted much excitement in 2009, when he proposed that QCD axions would form large condensates in the very early universe. His calculations suggested they would lead to ringlike galaxy halos rather than the spherical halos that most astronomers expect and that WIMP models predict. If so, then we might be able to tell what dark matter is made of just by looking at halo shapes.

But when Mark, Alan and I sat down to check how Sikivie's group had arrived at this prediction, we came to a radically different conclusion. Although we agreed that axion Bose-Einstein condensates would form in the early universe, they would be much smaller–the size of asteroids. Our model also did not give any indications, in the present-day universe, of what kind of axion structures we might find billions of years in the future. Trying to better model how–and whether–we get from small asteroid-sized condensates to the galactic-scale dark matter halos of today is still a significant computational challenge.

The same year our paper came out, another group was looking into other interesting implications of axionlike particles. A team led by Hsi-Yu Schive of National Taiwan University published computer simulations of certain axionlike particles that are often referred to as "ultralight axions" or "fuzzy dark matter," so named because they have a very low mass and would act like blurred-out waves rather than pointlike particles. They showed that these particles could form wavelike dark matter halos with Bose-Einstein condensates at their cores. Schive's paper generated new interest in ultralight axions and raised hopes that astrophysical observations could detect signs of the wavelike halo structures we expect.

These days axions and axionlike particles stand along with WIMPs as some of our best guesses at what dark matter could be. Another category that is growing in popularity is a model called self-interacting dark matter (SIDM). This idea predicts fermion dark matter particles that have some kind of interaction with one another–a self-interaction–beyond gravity. These self-interactions could create more interesting shapes and structures within a halo than a smooth, spherical blob. The particulars of the structures are hard to predict, though, and depend on the mass and other characteristics of the particles. Interestingly, axions may also interact with one another, though in different ways than self-interacting fermions.

There is an alternative to WIMPs, axions and SIDM: neutrinos. Although Standard Model neutrinos are now known to be too low in mass to explain all of the missing matter, these neutrinos are real and hard to see, making them functionally a small component of the dark matter that we call the cosmic neutrino background. In addition, a new type of neutrino has been hypothesized as a companion to the Standard Model neutrino: the sterile neutrino. Sterile neutrinos are distinct because they interact primarily gravitationally and only mildly through Standard Model forces. In addition, they are perhaps the most popular warm—or at least somewhere between hot and cold—dark matter proposal.

Another idea that theorists are just starting to explore is that rather than a single dark matter particle, there may be an entire sector. Perhaps dark matter is made of traditional axions, axionlike particles, WIMPs, sterile neutrinos and SIDM—all together. One other tantalizing possibility is that dark matter actually comprises stellar-mass black holes that would have formed in the early universe. This option has become more popular since the 2017 detection of gravitational waves indicated that black holes in this mass range are more common than expected.

Clues in the Sky

In astronomy we are relatively passive observers. We can choose our instruments, but we cannot design a galaxy or a stellar process and watch it unfold. Cosmic phenomena rarely happen on human-friendly time scales—galaxy formation takes billions of years, and the cosmic processes that might emit dark matter particles do so over tens to hundreds of years.

Even so, astrophysical probes of dark matter can tell us a lot. For instance, the NASA Fermi Gamma-ray Space Telescope has functioned as a dark matter experiment by looking for gamma-ray signatures that could be explained only by dark matter. WIMPs, for instance, are predicted to be their own antimatter partners, meaning that if two WIMPs collided, they would annihilate each other just as matter and antimatter do on contact. These explosions should produce an abundance of gamma-ray light where there is dark matter, especially at the cores of galaxies where dark matter is densest.

In fact, the Fermi telescope does see an excess of gamma-ray light at the center of the Milky Way. These observations have inspired passionate debate among observers and theorists. One interpretation is that these fireworks result from dark matter colliding with itself. Another possibility is that the signal comes from neutron stars near the center of the Milky Way that emit gamma-ray light through the typical course of their lives. Some astrophysicists favor the more mundane neutron star explanation, but others think the signal is dark matter. The fact that there is disagreement is normal, and even I have a hard time deciding what I think. I am compelled by physicists Tracy Slatyer and Rebecca Leane's thoughtful research showing that a dark matter explanation is sensible, but in the end, only analysis of more detailed observations will persuade the community about either idea. Future data from the Fermi telescope and proposed experiments such as NASA's All-sky Medium Energy Gamma-ray Observatory eXplorer (AMEGO-X for short) have the potential to settle the debate.

Scientists have also used the Fermi telescope to look for evidence of axions. Theories predict that when axions encounter magnetic fields, they occasionally decay into photons. We hope that by looking over long distances, we might see signs of this light, offering proof that axions exist. And neutron stars–the potential confounding signal at the Milky Way's center–are actually a good place to look for dark matter on their own. Some theories suggest that these dense spinning stars produce axions when protons and neutrons collide in their cores. We might be able to observe these axions as they decay into photons and escape from the stars. And as neutron stars release the dark matter over tens to hundreds of years, they would cool down in a pattern that we may be able to measure–if we look long enough. Another hot topic of study right now is whether nonaxion dark matter collects in neutron stars, affecting the structure of the star. As a member of the NASA Neutron Star Interior Composition Explorer (NICER) collaboration, I am leading a research project that is using data from NICER, a little telescope on the International Space Station that is up for renewal later this year. Our project is looking for evidence that dark matter is inside or enveloping neutron stars.

We can also learn more about the nature of dark matter by studying the best evidence we have for its existence so far–the cosmic microwave background (CMB) radiation. This light is a radio signal that originated in the early universe, and it is inescapably everywhere, all around us. It provides a snapshot of a moment early in cosmic history, and the patterns we see in the frequencies of its light reflect the makeup of the universe when it was created. It turns out that we can only explain the patterns we see in the CMB by assuming that dark matter was present–if there were no dark matter, the CMB data would make no sense. The patterns in the data tell us what fraction of the total mass and energy dark matter contributed; they even help constrain the possible masses of the dark matter particles. As I write, the CMB-Stage 4 collaboration is preparing to use a collection of telescopes in Chile's Atacama Desert and at the South Pole to take the most detailed measurements yet of the CMB.

On the Horizon

Since that 2009 Women in Astronomy conference, Rubin and Roman have both passed away, but their legacies live on through projects that will seek to better understand our universe. The NASA Nancy Grace Roman Space Telescope will launch in the mid-2020s, and although it is primarily focused on studying cosmic acceleration (the "dark energy problem") and exoplanets, it will also provide insight into dark matter. At the same time, here on Earth, the Vera C. Rubin Observatory in the Atacama Desert will support research on many questions, including the search for the dark matter that made Rubin famous.

In other words, we have lots to look forward to in the coming years. One reason is that almost any large-scale astronomical observation has something to tell us about dark matter. For example, a team in Mexico led in part by Alma X. Gonzalez-Morales and Luis Arturo Ureña-López showed that we can use the phenomenon of gravitational lensing, where large masses bend spacetime so much that it acts like a fun-house mirror, to place constraints on the mass of fuzzy dark matter. Gonzalez-Morales and Ureña-López are both active participants in the Rubin Observatory's Legacy Survey of Space and Time program, working on gravitational lensing and participating in the dark matter working group. Within the group, we are discussing observations that will capture more detailed information about dark matter halos that can then be compared with computer simulations of proposed dark matter candidates. Similarly,

Roman telescope surveys of large-scale structure will provide insight into dark matter's behavior on cosmic scales.

In the future, proposed x-ray observatories such as the NASA Spectroscopic Time-Resolving Observatory for Broadband Energy X-rays (STROBE-X) can help us take a closer look at neutron star structure in ways that will enhance our understanding of dark matter's possible properties. Other proposed future projects such as NASA's All-sky Medium Energy Gamma-ray Observatory, or AMEGO (not to be mistaken for AMEGO-X), will do the same in a different wavelength.

I will be an active participant not just as a scientist but as one of three conveners, alongside Alex Drlica-Wagner and Hai-Bo Yu, for the Snowmass Cosmic Frontier's topic Dark Matter: Cosmic Probes. It is our responsibility to describe the excitement and possibilities of astrophysical searches for dark matter to the funding decision-makers. The document I will help produce may influence guidance that is given to the National Science Foundation and the U.S. Department of Energy about what research we conduct over the next decade.

Coincidentally, the astronomy community just recently completed a similar process known as the 2020 Decadal Survey on Astronomy and Astrophysics. The resulting report sidestepped substantively addressing the dark matter problem, but it still offered strong support for efforts to better map the CMB, instruments to study neutron stars and x-ray observatories–three goals that will help us understand dark matter.

Doing science is never just about calculations, observations and experiments; it is also about working collaboratively with other people, including policy makers. How much progress we make will depend in part on what kind of support we get from lawmakers. Of course, this is stressful to think about. The good news is that there is a universe to wonder about, and trying to understand dark matter is a great distraction.

About the Author

Chanda Prescod-Weinstein is a theoretical physicist. She is an assistant professor of physics and a core faculty member in women's and gender studies at the University of New Hampshire. She is author of The Disordered Cosmos: A Journey into Dark Matter, Spacetime, and Dreams Deferred *(Bold Type Books, 2021).*

How Analyzing Cosmic Nothing Might Explain Everything

By Michael D. Lemonick

Computational astrophysicist Alice Pisani put on a virtual-reality headset and stared out into the void—or rather a void, one of many large, empty spaces that pepper the cosmos. "It was absolutely amazing," Pisani recalls. At first, hovering in the air in front of her was a jumble of shining dots, each representing a galaxy. When Pisani walked into the jumble, she found herself inside a large swath of nothing with a shell of galaxies surrounding it. The image wasn't just a guess at what a cosmic void might look like; it was Pisani's own data made manifest. "I was completely surprised," she says. "It was just so cool."

The visualization, made in 2022, was a special project by Bonny Yue Wang, then a computer science undergraduate at the Cooper Union for the Advancement of Science and Art in New York City. Pisani teaches a course there in cosmology—the structure and evolution of the universe. Wang had been aiming to use Pisani's data on voids, which can stretch from tens to hundreds of millions of light-years across, to create an augmented-reality view of these surprising features of the cosmos.

The project would have been impossible a decade ago, when Pisani was starting out in the field. Scientists have known since the 1980s that these fields of nothing exist, but inadequate observational data and insufficient computing power kept them from being the focus of serious research. Lately, though, the field has made tremendous progress, and Pisani has been helping to bring it into the scientific mainstream. Within just a few years, she and an increasing number of scientists are convinced, the study of the universe's empty spaces could offer important clues to help solve the mysteries of dark matter, dark energy and the nature of the enigmatic subatomic particles called neutrinos. Voids have even shown that Einstein's general theory of relativity probably operates the same way at very large scales as it does locally—something that has never been confirmed. "Now is the right moment to use voids" for cosmology, says David Spergel, former chair of astrophysics at Princeton University and current president of the Simons Foundation. Benjamin Wandelt of the Lagrange Institute in Paris echoes the sentiment: "Voids have really taken off. They're becoming kind of a hot topic."

The discovery of cosmic voids in the late 1970s to mid-1980s came as something of a shock to astronomers, who were startled to learn that the universe didn't look the way they'd always thought. They knew that stars were gathered into galaxies and that galaxies often clumped together into clusters of dozens or even hundreds. But if you zoomed out far enough, they figured, this clumpiness would even out: at the largest scales the cosmos would look homogeneous. It wasn't just an assumption. The so-called cosmic microwave background (CMB)—electromagnetic radiation emitted about 380,000 years after the big bang—is extremely homogeneous, reflecting smoothness in the distribution of matter when it was created. And even though that was nearly 14 billion years ago, the modern universe should presumably reflect that structure.

But we can't tell whether that's the case just by looking up. The night sky appears two-dimensional even through a telescope. To confirm the presumption of homogeneity, astronomers needed to know not only how galaxies are distributed across the sky but how they're distributed in the third dimension of space—depth. So they needed to measure the distance from Earth to many galaxies near and far to figure out what's in the foreground, what's in the background and what's in the middle. In 1978 Laird A. Thompson of the University of Illinois Urbana-Champaign and Stephen A. Gregory of the University of New Mexico did just that and discovered the first hints of cosmic voids, shaking the presumption that the universe was smooth. In 1981 Harvard University's Robert Kirshner and four of his colleagues discovered a huge void, about 400 million light-years across, in the direction of the constellation Boötes. It was so big and so empty that "if the Milky Way had been in the center of the

Boötes void, we wouldn't have known there were other galaxies [in the universe] until the 1960s," as Gregory Scott Aldering, now at Lawrence Berkeley National Laboratory, once put it.

In 1986 Margaret J. Geller, John Huchra and Valérie de Lapparent, all then at Harvard, confirmed that the voids Thompson, Kirshner and their colleagues had found were no flukes. The team had painstakingly surveyed the distance to many hundreds of galaxies spread out over a wide swath of sky and found that voids appeared to be everywhere. "It was so exciting," says de Lapparent, now director of research at the Institut d'Astrophysique de Paris (IAP). She had been a graduate student at the time and was spending a year working with Geller, who was trying to understand the large-scale structure of the universe. A cross section of the local cosmos that astronomers had put together earlier showed hints of a filamentary structure consisting of regions either overdense or underdense with galaxies. "Margaret had this impression that this was just an observing bias," de Lapparent says, "but we had to check. We wanted to look farther out." They used a relatively small telescope on Mount Hopkins in Arizona. "I learned to observe on that telescope," de Lapparent recalls. "I was on my own after a night of training, which was so exciting." When she was done, she, Geller and Huchra made a map of the galaxies' locations. "It was amazing," she says. "We had these big, circular voids and these sharp walls full of galaxies."

"All of these features," the researchers wrote in their paper, entitled "A Slice of the Universe," "pose serious challenges for current models for the formation of large-scale structure." As later, deeper surveys would confirm, galaxies and clusters of galaxies are themselves concentrated into a gigantic web of concentrated regions of matter connected by streaming filaments, with gargantuan voids in between. In other words, the cosmos today vaguely resembles Swiss cheese, whereas the CMB looks more like cream cheese.

The question, then, was: What forces made the universe evolve from cream cheese into Swiss cheese? One factor was almost certainly dark matter, the invisible mass whose existence had in the 1980s only recently been accepted by most astrophysicists, despite years of tantalizing evidence from observers such as Vera Rubin and Fritz Zwicky. It was more massive than ordinary, visible matter by a factor of six or so. That would have made the gravitational pull of slightly overdense regions in the early universe stronger than anyone had guessed. Stars and galaxies would have formed preferentially in these areas of high density, leaving low-density regions largely empty.

Most observers and theorists continued to explore what would come to be known as the "cosmic web," but very few concentrated on voids. It wasn't for lack of interest; the problem was that there wasn't much to look at. Voids were important not because of what they contained but because their very existence, their shapes and sizes and distances from one another, had to be the result of the same forces that gave structure to the universe. To use voids to understand how those forces worked, astrophysicists needed to include many examples in statistical analyses of voids' average size and shape and separation, yet too few had been found to draw useful conclusions from them. It was analogous to the situation with exoplanets in the 1990s: the first few discovered were proof that planets did indeed orbit stars beyond the sun, but it wasn't until the Kepler space telescope began raking them in by the thousands after its 2009 launch that planetary scientists could say anything meaningful about how many and what kinds of planets populated the Milky Way.

Another issue with studying voids was raised in 1995 by Barbara Ryden of the Ohio State University and Adrian L. Melott of the University of Kansas. Galaxy surveys, they pointed out, are conducted in "redshift space," not actual space. To understand what they meant, consider that as the universe expands, light waves are stretched from their original wavelengths and colors into longer, redder wavelengths. The farther away something is from an observer, the more its light is stretched. The James Webb Space Telescope was designed to be sensitive to infrared light in part so it can see the very earliest galaxies, whose light has been stretched all the way out of the visible spectrum—it's redder than red. And the CMB, the most distant light we can detect, has been stretched so much that we now perceive it in the form of microwaves. "Measuring the physical distances to galaxies is difficult," Ryden and Melott wrote in a paper in the *Astrophysical Journal*. "It's much easier to measure redshifts." But, they noted, redshifts can distort the actual distances to galaxies that enclose a void and thus give a misleading idea of their size and shape. The problem, explains Nico Hamaus of the Ludwig Maximilian University of Munich, is that as a void expands, "the near side is coming toward us, and the far side is streaming away." That differential subtracts from the redshift on the near side and adds to it on the far side, making the void look artificially elongated.

Despite the difficulties, astrophysicists began to feel more equipped to tackle voids by the late 2000s. Projects such as the Sloan Digital Sky Survey had probed much more deeply into the cosmos than the map made by Geller, Huchra and de Lapparent and confirmed that voids were everywhere you looked. Independent observations by two teams of astrophysicists, meanwhile, had revealed the existence of dark energy, a kind of negative gravity that was forcing the universe to expand faster and faster rather than slowing down from the mutual gravitational attraction of trillions of galaxies. Voids seemed to offer astronomers a promising way of studying what might be driving dark energy.

These developments caught Wandelt's eye. His specialty has always been trying to understand how the large-scale structure of the modern universe came to be. One of the aspects of voids that he found attractive, he says, was that "these underdense regions are much quieter in some ways, more amenable to modeling" than the clusters and filaments that separate them. Galaxies and gases are crashing into each other in nonlinear and complicated interactions, Wandelt says. There's "a chaos" that erases the information about their formation. Further complicating things, the gravitational attraction between galaxies is strong enough on smaller scales that it counteracts the general expansion of the universe–and even counteracts the extra oomph of dark energy. Andromeda, for example, the nearest large galaxy to our own, is actually drawing closer to the Milky Way; in four billion years or so, they'll merge. Voids, in contrast, "are dominated by dark energy," Wandelt says. "The biggest ones are actually expanding faster than the rest of the universe." That makes them ideal laboratories for getting a handle on this still puzzling force.

And it's not just an understanding of dark energy that could emerge from this line of study; voids could also cast light (so to speak) on the nature of dark matter. Although voids have much less dark matter in them than the clusters and filaments of the cosmic web do, there's still some. And unlike the chaotic web, with its swirling hot gases and colliding galaxies, the voids are calm enough that the particles astrophysicists think make up dark matter might be detectable. They wouldn't show up directly, because they neither absorb nor emit light. But the particles should occasionally collide, resulting in tiny bursts of gamma rays. They would also probably decay eventually, releasing gamma rays in that process as well. A sufficiently sensitive gamma-ray telescope in space would theoretically be able to detect their collective signal. Nicolao Fornengo of the University of Turin in Italy, co-author of a preprint study laying out this rationale, says that "if dark matter produces [gamma rays], the signal should be in there."

Voids could even help to nail down the nature of neutrinos—elementary particles, once thought to be massless, that pervade the universe while barely interacting with ordinary matter. (If you sent a beam of neutrinos through a slab of lead one light-year, or nearly six trillion miles, thick, about half of them would sail through it effortlessly.) Physicists have confirmed that the three known types of neutrinos do have masses, but they aren't sure why or exactly what those masses are.

Voids could help them find the answer, says Elena Massara, a postdoctoral researcher at the Waterloo Center for Astrophysics at the University of Waterloo in Canada. They're places that have a lack of both luminous matter and dark matter, she explains, "but they're full of neutrinos, which are almost uniformly distributed" through the universe, including in voids.That's because neutrinos zip through the cosmos at nearly the speed of light, which means they don't clump together under their mutual gravity—or under the gravity of the dark matter concentrations that act as the scaffolding for the cosmic web.

Although voids always contain a lot of neutrinos, the particles are only passing through—those that fly out are constantly replenished by more neutrinos streaming in. And their combined gravity can make the voids grow more slowly over time than they would otherwise. The rate of growth—determined through comparison of the average size of voids in the early universe to those in the modern universe—can reveal how much mass neutrinos actually have.

Void science has changed a lot since Pisani started studying it as a graduate student working with Wandelt. He offered two or three suggestions for a dissertation topic, she recalls, and one of them was cosmic voids. "I felt that they were the riskiest choice," she says, "because there were very few data at the time. But they were also incredibly challenging," which she found exciting. The data Pisani and others needed to analyze the voids, however—that is, to test their real-world properties against computer models incorporating dark matter, dark energy, neutrinos and the formation of large-scale structure in the universe—were simply not available. "When I started my Ph.D. thesis," Pisani says, "we knew of fewer than 300 voids, something like that. Today we have on the order of 6,000 or more."

That's huge, but it's still not enough for the comprehensive statistical analysis necessary for voids to be used for serious cosmology—with one exception. In 2020 Hamaus, Pisani, Wandelt and several of their colleagues published an analysis showing that general relativity behaves at least approximately the same way on very large scales as it seems to do in the local universe. Voids can be used to test this question because astrophysicists think they result from the way dark matter clusters in the universe: the dark matter pulls in ordinary matter, creating the cosmic web and leaving empty spaces behind. But what if general relativity, our best theory of gravity, breaks down somehow over very large distances? Few scientists expect that to be the case, but it has been suggested as a means to explain away the existence of dark matter.

By looking at the thickness of the walls of matter surrounding voids, however, Hamaus and his colleagues determined that Einstein's theory is safe to rely on. To understand why, imagine a void as "a circle whose radius increases with the expansion of the universe," Wandelt says. As the circle grows, it pushes against the boundaries of galaxies and clusters at its perimeter. Over time these structures aggregate, thickening the "wall" that defines the void's edge. Dark energy and neutrinos affect the thickness as well, but because they are smoothly distributed both inside and outside the voids, they have a much smaller effect overall.

Scientists plan to use voids to learn even more about the universe soon because they expect to rapidly multiply the number of known voids in their catalog. "In the next five or 10 years," Pisani says, "we're going to have hundreds of thousands. It's one of those fields where numbers really make a difference." So, Spergel says, do advances in machine learning, which will make it far easier to analyze void properties.

These exploding numbers won't be coming from projects explicitly designed to search for voids. They will arrive, as they did with the Sloan Digital Sky Survey, as a by-product of more general surveys. The European Space Agency's Euclid mission, for example, which launched in July 2023, will create a 3-D map of the cosmic web with unprecedented breadth and depth. NASA's Nancy Grace Roman Space Telescope will begin its own survey in 2026, looking in infrared light. And in 2024 the ground-based Vera C. Rubin Observatory will launch a 10-year study of cosmic structure, among other things. Combined, these projects should increase the inventory of known voids by two orders of magnitude.

"I remember one of the first talks I gave on void cosmology, at a conference in Italy," Pisani says. "At the end the audience had no questions." She wasn't sure at the time whether the reason was skepticism or simply that the topic was so new to her listeners that they couldn't think of anything to ask. In retrospect, she thinks it was a little of both. "Initially, I think the problem was just convincing people that this was reasonable science to look into," she says.

That's much less of an issue now. For example, Pisani points out, the Euclid voids group has about 100 scientists in it. "I have to say that Alice was one of the fearless pioneers of this field,"

Wandelt notes about his former Ph.D. student. When they started writing the first papers on void science, he recalls, some of the leading figures in astrophysics "expressed severe doubt that you could do anything cosmologically interesting with voids." The biggest confirmation that they were wrong, he says, is that some of those same people are now enthusiastic.

Pisani is perhaps the ideal representative for this fast-emerging field. She approaches the topic with absolute scientific rigor but also with infectious enthusiasm. Whenever she talks about voids, she lights up, speaking rapidly, jumping to her feet to draw diagrams on a whiteboard, and fielding questions (of which there are now many) with ease and confidence. She emphasizes that void science won't answer all of astrophysicists' big questions about the universe by itself. But it could do something even more valuable in a way: test ideas about dark matter, dark energy, neutrinos and the growth of cosmic structure independently of the other strategies scientists use. If the results match, great. If not, astrophysicists will have to reconcile their differences to find out what's actually going on in the cosmos.

"I find the idea attractive and even somewhat poetic," Wandelt says, "that looking into these areas where there's nothing might yield information about some of the outstanding mysteries of the universe."

About the Author

Michael D. Lemonick is a freelance writer, as well as former chief opinion editor at Scientific American *and a former senior science writer at* Time. *His most recent book is* The Perpetual Now: A Story of Amnesia, Memory and Love *(Doubleday, 2017). Lemonick also teaches science journalism at Princeton University.*

The Future of Gravitational Wave Astronomy

By Lee Billings

A century ago, when Albert Einstein first predicted the existence of gravitational waves—subtle ripples in spacetime produced by massive objects hurtling through the cosmos—he also guessed they could not ever be seen. Although the echoes of distant celestial symphonies must ripple through the very fabric of reality, Einstein thought their ethereal harmonies were destined to remain eternally unheard.

On Thursday, scientists using the Laser Interferometer Gravitational-Wave Observatory (LIGO) proved Einstein both right and wrong, announcing their detection of the first note in a cosmic symphony he predicted no one would ever hear. It was a burbling chirp of gravitational waves produced by the cataclysmic birth of a black hole from the merger of two smaller ones. Emitted in a distant galaxy when multicellular life was just beginning to populate Earth, the waves traveled at the speed of light for more than a billion years to at last wash over our planet last September, taking just seven milliseconds to traverse the distance between LIGO's twin listening stations in Louisiana and Washington State.

Now, unlike Einstein a century ago who could scarcely imagine gravitational waves ever being seen, the scientists hunting the elusive spacetime ripples already have big plans for more detectors and observatories in the near and far future.

"Imagine light having never been collected in a photograph," says Janna Levin, an astrophysicist at Barnard College of Columbia University and author of a forthcoming book about LIGO. "The first thing people want to do is just to capture the recording, which is what LIGO has done."

Soon, astronomers say, LIGO will record and unveil far more than the birth cries of newborn black holes. It and other operational observatories are already looking for ripples from the violent death throes of massive stars and from collisions of city-size orbs of degenerate matter called neutron stars. Current observatories could also help reveal what makes spinning neutron stars called pulsars tick, mapping their starquake-shaken interiors and any centimeters-high "mountains" (which would weigh roughly the mass of a planet because of neutron stars' extreme density) that could pop up on their surfaces.

Decades from now new generations of space telescopes could capture the mergers of supermassive black holes and glimpse pulsars spiraling to doom down their maws, or see snapping "cosmic strings," proton-thin intergalactic defects in spacetime that may have been stretched across the infant universe during an inflationary growth spurt. Tracked and timed by radio telescopes, rapidly spinning pulsars can themselves be transformed into galaxy-spanning detectors sensitive to spacetime ripples with wavelengths measured in light-years. Ultimately, the most ambitious gravitational wave observatories astronomers can presently conceive might someday record the hiss of waves emitted in the first fractions of a trillionth of a second after the big bang. Then, cosmologists could watch–could listen–as the first seeds of cosmic structure crystallized from a seething quantum fog.

What most excites scientists, though, is the unknown. "Are there things out there that we've never even wrapped our heads around with telescopes?" Levin wonders. "Seeing black holes collide is a golden discovery, but we expected that. What else is out there? I want to see something dark."

"The skies will never be the same," says Szabolcs Márka, a physicist and LIGO team member at Columbia University. "Imagine you can touch, smell, taste, and see–and one day you can hear. That day is a glorious day. This is what has happened to us, as humanity. From today, we can hear the cosmos. We can see the unseen."

The Most Expensive "Nothing" Ever Made

Having found its first signal, LIGO is now gearing up to transform them into routine tools for astronomy. The twin LIGO stations each pass laser light back and forth between mirrors along perpendicular four-kilometer-long arms arranged in an L. An incoming wave would slightly warp these arms so that one became longer or shorter than the other by only a few thousandths the radius of a single proton, altering the flight time of the light and triggering a detection. Any number of background noises can scuttle the delicate measurement–LIGO can also hear ocean waves pounding distant coastlines, airplanes flying overhead and even the seismic hum from washing machines.

"We're trying to detect something smaller than the atoms our detector is built of," says Imre Bartos, a LIGO member and lecturer at Columbia University. "It doesn't sound believable, to be honest with you." That LIGO can hear gravitational waves at all over the cacophony of background noise is due to both stations recently receiving a series of noise-canceling "Advanced LIGO" upgrades that will soon make them 10 times more sensitive than they were during a fruitless first-generation search that occurred between 2002 and 2010. All told, the upgrades bring the experiment's total cost to more than $1 billion, most of it paid by the National Science Foundation. LIGO's ultra-high-vacuum hermetic perfection, Márka quips, is "the most expensive 'nothing' ever made."

As Advanced LIGO reaches its maximum sensitivity and plans a third listening station in India, it will work in tandem with other European laser interferometers such as GEO600 and Advanced VIRGO to rapidly make the detection of gravitational waves routine. "We've just made a machine that has given humanity a new sense, beyond the usual five," says LIGO team member Rana Adhikari, a Caltech physicist who helped develop the upgrades. "We're going to have to learn what it is like to feel the burbling of space with these brand-new gravitational fingers."

With gravitational waves set to soon ripple through several high-precision laser interferometers on Earth, astronomers will be also be able to locate where exactly each set of the ripples is coming from. LIGO's first detection of colliding black holes, by contrast, could only be traced to a huge arc of sky over the Southern Hemisphere. Pinpointing the sources of gravitational waves will allow astronomers to point other telescopes their way, boosting the chances of learning more about them via x-rays, gamma-rays, radio waves, neutrinos and more.

Burying the Noise

The current generation of ground-based laser interferometers can only go so far, however. The length, number and location of an interferometer's arms intimately influence its resilience against background noise and the varieties of gravitational waves it can probe. LIGO's four-kilometer-long arms are the biggest in the world, but at the project's Louisiana site the encroaching, noisy sprawl of nearby Baton Rouge is getting too close for comfort to the delicate detectors.

Researchers are now planning and building a next generation of even bigger and more isolated detectors deep beneath the ground where hundreds of meters of overlying rock shield against most anthropogenic noises and seismic stresses. In the Kamioka mine in Japan, the Kamioka Gravitational Wave Detector (KAGRA) is already taking shape as workers construct twin sets of three-kilometer arms in newly bored tunnels. Slated to enter operation in 2018, KAGRA will use cryogenically cooled mirrors of sapphire to deliver LIGO-like sensitivity.

After KAGRA, a consortium of European partners is forming tentative plans for an even more ambitious subterranean laser interferometer, the Einstein Telescope, which could come online in the late 2020s at a cost of $1 billion or $2 billion. Although it presently lacks funding and a construction site, its conceptual design calls for dual cryogenic and room-temperature beam lines running through three 10-kilometer arms arranged in the shape of an equilateral triangle rather than an L. That configuration would help it pinpoint the sources of gravitational waves on the sky and allow it to see the longer-wavelength ripples from a wider range of sources including binary white dwarfs, slower-spinning pulsars and intermediate-mass black holes weighing hundreds or thousands of suns. It would also begin to construct a reasonably detailed map of "background" sources of gravitational waves—the accumulated ripples from all the messy, violent mergers and explosions all across the sky.

"People wonder why we are not content with one gravitational-wave detector, why we wish to build bigger ones," says Harald Lück, a physicist at the Max Planck Institute for Gravitational Physics in Hannover, Germany who is a member of the GEO600 and Einstein Telescope teams. "Like electromagnetic radiation, gravitational waves cover an incredibly large range of wavelengths, and you can't catch all of them with any single facility." On the ground, Lück says, the arms of laser interferometers are unlikely to ever exceed 50 kilometers—past that, seismic noise, Earth's curvature, imperfect optics and the great expense of digging deep tunnels would outweigh any conceivable scientific gains.

Sooner or later, it will simply be cheaper to forsake Earth, to build and operate truly giant gravitational-wave observatories in space.

Space Is the Place

Whenever the first gravitational wave mission launches into space, radio astronomers will wryly say they were there first. Using radio telescopes on the ground, researchers have already devised space-based gravitational wave detectors using nature's resources: large numbers of pulsars spread throughout space that spin once every few milliseconds, sending out lighthouse-like beams of light that reach us in regular beats. By cross-correlating the arrival times of all the different pulses to nanosecond precision across decades, astronomers hope to detect gravitational waves with wavelengths measured in light-months and light-years as their passing periodic ripples distort spacetime around Earth. Merging galaxies should produce such extra-large waves as supermassive black holes at their galactic cores lock into orbital pairs and eventually collide. As compelling as the technique is, it has yet to deliver any detections, although the likelihood of success grows over time as astronomers incorporate more pulsars into their observations.

In addition to using pulsar timing, many researchers hope to also study gravitational waves through space-based laser interferometers. Freely floating in deep space far from Earth's noise and gravity, such facilities could in theory boast arms of almost any length. In practice, however, designers of possible future missions have grappled with the great complexity of engineering such ambitious spacecraft as well as new frontiers of contaminating noise in space. Those limitations have constrained their designs to arm lengths of "only" millions of kilometers.

In the 2000s NASA and the European Space Agency jointly worked to develop a Laser Interferometer Space Antenna (LISA), a constellation of three satellites that would form a triangular interferometer with three five-million-kilometer arms. That would be big enough to see gravitational waves emitted by any merging supermassive black holes that may have existed around the time when the universe's first stars began to shine, about a hundred million years after the big bang.

In 2011, however, NASA abandoned the effort due to its high estimated cost and shortfalls in the agency's science funding. ESA regrouped and downsized the project to become "eLISA," an "evolved" design calling for three satellites forming one-million-kilometer arms that could lift off in the mid-2030s. Last December ESA launched eLISA's precursor, a technology development mission called LISA Pathfinder. Meanwhile, other researchers are attempting to develop an alternate space-based interferometry technology that uses microscopic clouds of atoms rather than bulky spacecraft to further shrink the need for long, expensive baselines.

Regardless of the interferometric technique used, "the science case for a space-based interferometer has never been in doubt," says Paul McNamara, ESA's project scientist for LISA Pathfinder. "It would allow us to look at the biggest, most violent events in the universe and map the entire history of mergers across a huge slice of cosmic time.... What has always been in doubt is whether it is possible to build an instrument that can measure to picometer-precision across millions of kilometers after having the life shaken out of it by a rocket launch."

Currently under commissioning in an orbit some 1.5 million kilometers sunward of Earth, LISA Pathfinder is set to begin science operations in March, although it will not be capable of seeking gravitational waves. Instead, it will demonstrate that its payload of two gold-platinum cubes can come as close as possible to a state of weightless, perfect rest. Held in a shielded vacuum enclosure within the spacecraft, the cubes should feel essentially no outside forces other than gravity. If separated by millions of kilometers and linked by lasers, such a system could then detect the exquisitely small distance changes caused by passing gravitational waves. Reaching a state of perfect rest within a sunlight-baked spacecraft studded with heaters, antennas, and thrusters is far from easy, however.

"Think of the weight of a bacterium sitting on your hand—that's roughly the same amount of residual acceleration, of force, that a cube [is allowed to] deviate from freefall" on the spacecraft, McNamara says. Lasers on LISA Pathfinder measure a cube's distance from the enclosure's walls, directing microthrusters that keep the spacecraft centered on the cube. The microthrusters are so minuscule that a thousand firing at full blast could barely lift a sheet of notebook paper in Earth's gravity—yet the 450-kilogram LISA Pathfinder only carries six.

Mapping the Newborn Cosmos

In 2005, before it abandoned LISA, NASA conducted a small study investigating a possible super-sized successor comprising a complex arrangement of four LISA-like constellations with beefier lasers and telescopes. Called the "Big Bang Observer," the mission as conceived would probably be so astonishingly expensive that few if any researchers have dared to actually estimate its price tag. It would be sensitive to nearly all the same targets investigated by LIGO and LISA out to distances of 10 billion light-years or more, but its true goal would be to assemble a detailed map of gravitational waves from the earliest eras of the primordial universe.

This "stochastic gravitational wave background" would contain crucial information about the universe's very first moments and its mysterious era of inflation, when it seems to have almost instantaneously ballooned from proton- to grapefruit-sized, amplifying quantum-scale density fluctuations along the way that shaped the first large-scale cosmic structures. In its importance for our understanding of–well, everything–measuring such a signal would be even more revolutionary than mapping the cosmic microwave background (CMB), the relic light from when the early universe first cooled to transparency some 380,000 years after the big bang.

The CMB is a literal firewall on efforts to peer further back in time using light–nearly all information about earlier events carried by photons was erased as they bounced aimlessly within the hot, dense plasma that suffused the infant universe. Yet for primordial gravitational waves, the CMB is not a firewall–it is a window. Those with very long wavelengths could leave faint imprints in the CMB to mark their passing. Most primordial waves would breeze through without a trace, however, scarcely interacting with anything else in the universe for all the rest of time–at least until they ripple through the right detector.

"In a way, the stochastic background is the hardest thing to detect, but also the one which would offer you the most insight, because black holes and neutron stars are kind of old hat," says Bruce Allen, a LIGO team member and director of the Max Planck Institute for Gravitational Physics. "Let's suppose we jump ahead a thousand years to when our highly advanced civilization has built observatories that cover the whole gravitational spectrum." Such tools could map the spectrum of primordial gravitational waves and perhaps distinguish "kinks and bumps" within it that reveal the earliest and most epochal milestones in the universe's evolution.

"Maybe it's not a thousand years away," Allen goes on. "Maybe it's just a century until something like the Big Bang Observer, and a century after that we'll build something broadband. What matters is that getting the stochastic background's spectrum is one of the last things humans will ever be able to do with studies of gravitational waves. Somewhere out there in the universe there's probably a civilization that has already done it. We just haven't—yet."

About the Author

Lee Billings is a science journalist specializing in astronomy, physics, planetary science, and spaceflight, and is a senior editor at Scientific American. *He is the author of a critically acclaimed book,* Five Billion Years of Solitude: the Search for Life Among the Stars, *which in 2014 won a Science Communication Award from the American Institute of Physics. In addition to his work for* Scientific American, *Billings's writing has appeared in the* New York Times, *the* Wall Street Journal, *the* Boston Globe, Wired, New Scientist, Popular Science, *and many other publications. A dynamic public speaker, Billings has given invited talks for NASA's Jet Propulsion Laboratory and Google, and has served as M.C. for events held by* National Geographic, *the Breakthrough Prize Foundation, Pioneer Works, and various other organizations. Billings joined* Scientific American *in 2014, and previously worked as a staff editor at* SEED *magazine. He holds a B.A. in journalism from the University of Minnesota.*

Section 5: Policy Behind the Astronomy

This Report Could Make or Break the Next 30 Years of U.S. Astronomy

By Lee Billings

Ask astronomers what question they most want to answer, and you will get scattered responses: How did the first stars, galaxies and black holes form? What is the nature of dark matter and dark energy? Are we alone?

Each question demands its own large telescope: no ultimate, one-size-fits-all instrument will ever exist, for none can be made to gather each and every kind of cosmic light. Black holes sometimes shine in x-rays, for instance, whereas Earth-like exoplanets are best studied in optical and infrared light. Yet such projects so strain the fraction of public and private funds allocated to astronomy that only a few—perhaps just one—can be prioritized at a time, leading to pileups of also-ran proposals and anxious researchers awaiting a rare chance to open new windows on the universe.

In the U.S., astronomers have managed these competing ambitions by devising a process that has become the envy of the scientific world: the Astronomy and Astrophysics Decadal Survey, a once-in-10-years exercise that recommends and ranks the community's priorities for the next decade—embodied, eventually, by major new federally-sponsored observatories on the ground and in space. Projects such as NASA's Hubble Space Telescope owe their existence, in part, to coveted endorsements from Decadals of yore, and the practice has spread to several other disciplines that now undertake Decadal Surveys of their own.

Organized by the National Academies of Sciences, Engineering, and Medicine, six Decadal Surveys have set the course of U.S. astronomy since they began in the 1960s. The results of the seventh, dubbed Astro2020, will soon be announced after two years of exhaustive deliberations led by a 20-member steering committee. And just like its predecessors, Astro2020 will reveal where major new investments and discoveries are most likely to be made—and where neglect, disinterest or even fear may block progress for generations to come.

Few people know the power of these surveys better than Joel Parriott. More than 20 years ago, he got his start in federal politics as a staffer at the National Academies, where his initial assignment was to assist the scientists crafting the first astronomy Decadal of the new millennium. Then he served a 10-year stint in the White House Office of Management and Budget (OMB), where he weighed and implemented Decadal recommendations for federal science agencies. Today he is director of public policy at the American Astronomical Society, the leading advocacy group for U.S. stargazers. At the OMB, where the nation's policy objectives often brutally intersect with its fiscal realities, the Decadal Survey offered Parriott and his colleagues a foolproof way of dismissing overly solicitous astronomers. "If a project wasn't highly ranked, we knew it didn't have the community's endorsement," he says. "That's really helpful for folks on Capitol Hill and in the White House who need to make hard choices."

For more than a few astronomers, a drab name like "the Decadal" does not properly capture a process that holds such sway over their destiny. Instead they sometimes just call it "the voice of God." In coming weeks, when Astro2020's final report is released, that voice—that supposedly *communal* voice—will once again speak. Yet outside of a chosen few, sworn to secrecy, no one in the community knows in advance what it will say. Everyone, though, agrees Astro2020's conclusions arrive at a time of peril.

"We are right now on a knife-edge," says John O'Meara, chief scientist of the W. M. Keck Observatory on Mauna Kea in Hawaii. "I do believe this Decadal is existential for astronomy in the United States. When you consider the facilities and the science topics that are under discussion, it will influence whether or not we become a second-place player in global astronomy.... When the [federal] agencies and Congress receive the Decadal report, they will hold in their hands the decision of whether or not we wish to have leadership in this field of science."

U.S. Telescopes in Twilight

From the outside looking in, one would not realize the enterprise of U.S. astronomy is teetering on the edge of crisis. Several new ground-based telescopes have recently come online, each bringing a bumper crop of celestial discoveries—and even more ambitious projects are waiting in the wings. Consider the Vera C. Rubin Observatory, a high-ranked priority of the past two Decadals (which are naturally called Astro2000 and Astro2010). Sited on a mountain in Chile, the Rubin Observatory should begin operations in late 2023 to generate a high-definition, decade-spanning time-lapse movie of the entire overhead sky.

But among large upcoming U.S. projects, Rubin is a rare, healthy exception. The other highest-profile ground-based recommendation from Astro2000 and Astro2010—building an optical extremely large telescope (ELT) with a mirror circa 30 meters in size—remains in limbo. One selling point for ELTs, among many, is that they offer the best hopes of ever studying Earth-like exoplanets from the ground. Astro2020 will probably decide if the U.S. ELT efforts sink or swim—or if other projects, such as a "next generation" upgrade to the nation's Very Large Array of radio telescopes or expansions of gravitational-wave observatories, take priority.

In the quest for an ELT, the U.S. has managed to produce two competing projects, the Thirty Meter Telescope (TMT) and the Giant Magellan Telescope (GMT). Both are short on funding, and neither appears likely to begin operations before the decade is out, each having served to stifle the other. And the TMT's early stages of construction on the Hawaiian volcano Mauna Kea—a site unrivaled for pristine views of the Northern Hemisphere sky—sparked protests from activists who see telescopes there as an occupying affront to the mountain, which Native Hawaiians hold sacred. Construction on the TMT ceased after protesters repeatedly blocked the road to the mountaintop; the conflict remains at an impasse. "If [Astro2020] says, 'Forget the ELTs; let's prioritize something else instead,' then it's quite possible that both the TMT and the GMT will die," says a senior ground-based astronomer familiar with the situation.

Europe, in contrast, took the lead over the U.S. in ground-based optical astronomy years ago and is well into construction of an ELT of its own in Chile. The European Extremely Large Telescope boasts a 40-meter mirror–and it is projected to come online in 2027.

China is surging ahead as well. For evidence, look no further than the U.S.'s iconic, National Science Foundation–funded Arecibo radio telescope in Puerto Rico: Once the world's largest, the radio telescope catastrophically collapsed last year in part because of budgetary neglect–but not before China's Five-Hundred-Meter Aperture Spherical Radio Telescope (FAST) had superseded it in size. And together–without the U.S.–Europe, China and many other international partners are building the Square Kilometer Array (SKA), a breathtakingly powerful collection of thousands of radio telescopes that is set to become fully operational at sites in Australia and South Africa as early as 2030.

The Hungry Giant and Frankenstein's Monster

The outlook is similarly mixed for the nation's space-based astronomy. For now, the U.S. remains at the forefront of off-world observing, but of the four "Great Observatories" NASA launched between 1990 and 2003, only Hubble and the Chandra X-ray Observatory are still operational, and both are nearing their end, with no replacement on the horizon. "Hubble is probably not going to last another decade, and maybe we'll get five more years out of Chandra. But then that's it–they're gone," says Jason Tumlinson, an astronomer heading the community missions office at the Space Telescope Science Institute. "We'll probably have a long gap with no real optical, ultraviolet or x-ray capability in space. And now is the time to decide how and when we might get it back."

Astro2000's top-ranked space project, NASA's flagship-class James Webb Space Telescope, is a technological marvel: a cryogenically cooled infrared observatory with a segmented, 6.5-meter starlight-gathering mirror that folds, origamilike, to fit in a rocket. Early on Webb was projected to cost about $1.5 billion and to launch perhaps in 2011 to study the emergence of stars and galaxies in the early universe. Today those projections seem hopelessly naive. After a staggering number of cost overruns and delays that hobbled planning for other projects, the current best-case scenario is that the telescope will reach space no sooner than this mid-November, operating for just a decade with a total project cost of about $10 billion. Less optimal scenarios, of which there are many, are almost too grim to contemplate. "We can't do science at this scale without taking risks–and I'm confident in our chances of success–but if Webb fails, it will be an unmitigated disaster," says Matt Mountain, the project's telescope scientist and president of the Association of Universities for Research in Astronomy. "It has to work. Because if it doesn't, we aren't going to do another ambitious flagship for, I would guess, two decades."

If Webb was a hungry giant unleashed by Astro2000 and its antecedents biting off more than could be chewed, then the top flagship recommendation of Astro2010, NASA's Nancy Grace Roman Space Telescope, was a different beast entirely–a cut-rate Frankenstein's monster the Decadal committee pieced together from the dismembered remains of multiple competing mission concepts. The committee had hoped to avoid another Webb-style debacle with Roman (initially named the Wide-Field Infrared Survey Telescope, or WFIRST)–and it did. But instead Roman's very existence threatened to tear the community apart from within. "Do you know what WFIRST really stood for, for most of us?" says one leading astronomer. "It stood for 'What the f*ck is this ridiculous space telescope?'"

Originally envisioned to study dark energy with a barebones instrument package and a mirror scarcely half the size of Hubble's, Roman was projected to launch as early as 2020 on a relatively slim budget of less than $2 billion. To many expert eyes, such a project barely qualified for its supposed "flagship" status. NASA, with bipartisan congressional blessings, ultimately added more instruments and upgraded Roman's mirror to the same size as Hubble's, enhancing its science objectives and assuaging many criticisms–but also nearly doubling its estimated price tag and delaying its launch to no earlier than 2025. Meanwhile Europe and China have each proceeded with dark energy-focused space telescopes of their own, potentially scooping some of the promised scientific discoveries used to justify Roman's existence in the first place.

Although both Webb and Roman may each eventually succeed beyond astronomers' wildest imaginings, some fear the projects' greatest initial impact on the field will be to sharply curtail Astro2020's ability to plan for a prosperous future. "This is, I think, the first Decadal where both of the top space-based recommendations from the previous two Decadals–Webb and Roman–were still on the ground," Tumlinson says. "And if the Astro2020 committee were to say, 'We've got two flagships stacked up; the queue is too long; let's just pause for a while and do some smaller missions and catch up later,' that would be a mistake. The idea that you can just take a decade off from ambitious things to come back and do them later is not valid when you consider how our government actually works."

Most astronomers, Tumlinson says, seem to misunderstand what the Decadal's "governing dynamic" actually is. "A Decadal report is the beginning of a multiparty, multiyear negotiation between the scientific community, NASA, the aerospace contractors, Congress and the White House," he explains–which is why aiming high at the outset is in astronomers' best interest. "I would hope, with Astro2020, we temper our natural desire to mitigate risk and cut costs," Tumlinson says, "because all the other forces in this system will be doing that for us anyway."

Whether because of the COVID-19 pandemic, soaring deficit spending or the increasingly dire global climate emergency, some might question the wisdom of U.S. astronomers reaching for the stars just as the sky seems set to fall. Then again, the reply comes, where is the wisdom in limiting the science of the 2040s or 2050s based on the troubles of the 2020s? "At the end of the day, appropriators are still going to spend money," says a former congressional staffer who dealt with high-level appropriations for federal science agencies. "They're going to get an allocation, and their job will be to use it wisely. If it doesn't go to astronomy, maybe it will go to a new flagship mission for NASA's planetary science division instead–or maybe it will go to a new FBI building. But that money will be spent, regardless of what astronomers do."

The Main Menu

Responding to the woes from Webb and Roman, years ago NASA began revamping its approach to future flagships, demanding greater certainty about technological challenges and costs. In 2016 the agency assembled four Science and Technology Definition Teams (STDTs), each examining a separate mission concept for Astro2020's consideration. Two of the four concepts–the Large Ultraviolet Optical Infrared Surveyor (LUVOIR) and the Habitable Exoplanet Observatory (HabEx)–would focus on the quest to learn more about planets orbiting other stars, with an emphasis on studying potentially habitable worlds. A third, the ultracold and far-infrared Origins Space Telescope, would also perform some exoplanet studies as part of a broader investigation of the formation of galaxies, stars and planetary systems. The fourth concept, the Lynx X-ray Observatory,would be the most powerful x-ray astronomy facility ever built, offering intimate views of black holes, active galaxies and violent supernovae across cosmic time. Each project has profound potential–but also one or more Achilles' heels to make astronomers antsy.

LUVOIR's strength–and weakness–has always been the enormity of its segmented mirror. If deployed and maintained with picometer-degree stability (which is very hard), this mirror would allow astronomers to discover and study hundreds of exoplanets while also performing revolutionary observations across a wide swath of general astrophysics. Assuming any true-blue Earth-like worlds exist around the sun's neighboring stars, LUVOIR should offer the best odds of finding them. But whether considering either a 15-meter "deluxe model" or an eight-meter "budget size" one, putting such a demanding deployable mirror into space translates to an astronomical cost. "When I started working on this, I closed my eyes and said, 'It's gonna be

$1 billion a meter. But if LUVOIR realizes its vision, that would be a bargain," says O'Meara, one of the leaders of the this mission concept's STDT. Estimates from two separate groups at NASA's Goddard Space Flight Center have arrived at slightly higher figures, placing a deluxe LUVOIR somewhere in the realm of $15 billion or $20 billion and finding the budget version between about $12 billion or $15 billion.

Thanks to its smaller four-meter mirror, HabEx would be cheaper than either version of LUVOIR, with an estimated cost approaching $10 billion. But it would yield far fewer exoplanets–delivering details on perhaps 10 potential exo-Earths rather than dozens. It represents an all-in bet on a novel technology: a secondary sunflower-shaped spacecraft called a "starshade" that would unfurl to more than 50 meters across and fly more than 75,000 kilometers in front of the four-meter mirror to blot out a target star's light, revealing comparatively dimmer accompanying worlds. "The starshade concept, although potentially very powerful, scares a lot of people for whom it is a relatively new idea," says Scott Gaudi, an astronomer at the Ohio State University, who co-chaired the HabEx STDT. Yet he maintains that scientists have a realistic plan for managing risks and developing the starshade on budget.

Origins is, in most respects, the "safe" choice: it features a large but nonsegmented mirror of nearly six meters and is based mostly on preexisting technologies, netting a cost estimate of around $7 billion. Chilled to less than five kelvins above absolute zero, the telescope would offer a 1,000-fold increase in far-infrared sensitivity over previous missions, allowing astronomers to map the inner workings of galaxies across the observable universe while also studying a handful of small exoplanets as well as water in protoplanetary disks around nearby stars. But Origins' tried-and-true approach makes it relatively bland: It would not answer the burning questions about Earth-like worlds that LUVOIR or HabEx might. Its infrared optics would not deliver the crisp, colorful images of Hubble. And a top ranking from Astro2020 would make it, after Webb and Roman, the third infrared flagship in a row recommended to NASA. Some might call Origins a space telescope only an astronomer could love. "We have a real PR problem in the infrared," says Cara Battersby, an astronomer at the University of Connecticut, who served on the Origins STDT. "But if you look at the specs of each [STDT] concept and the science questions they have in common, such as planets and the coevolution of black holes with galaxies, I challenge you to not conclude that Origins is the most well-rounded and safest of them all."

Lynx is the oddball of the four, with a projected price tag slightly shy of that of Origins but a radically different design and goal. Its "mirror" would be a first-of-its-kind three-meter-wide assembly of nearly 460 nested shells of polished silicon, all densely packed and angled to reflect and focus high-energy x-rays. That design would provide far better performance than Chandra or other earlier x-ray telescopes, allowing Lynx to unveil new details of the universe's oldest and biggest black holes. Some astronomers fret about potential budget-busting difficulties in making Lynx's exotic mirror, but most instead worry about duplicating the efforts of a similar, already approved project: the European Space Agency's Advanced Telescope for High-Energy Astrophysics (Athena), which is set to launch in 2031. "For many people, the existence of Athena is an insurmountable argument against Lynx," says Grant Tremblay, an astronomer and a contributor to the Lynx STDT at the Center for Astrophysics at Harvard University and the Smithsonian Institution."But I sincerely believe there is a very compelling argument for why these two missions are scientifically complementary to each other."

Collectively, these four STDT concepts make up the main menu of flagship options that the Astro2020 steering committee is most likely to choose from—presuming, that is, the committee picks one at all. All four projects are subject to Astro2020's Technical Risk and Cost Evaluation (TRACE) process, a brand-new, behind-closed-doors checkup by the Aerospace Corporation on each STDT concept's estimates. If the TRACE deems all four concepts far more expensive than the STDTs' in-house estimates, the steering committee could opt to choose none.

The New Great Observatories

Astronomers have named the four likeliest outcomes of the Astro2020 deliberations: "Scenario one, we call 'the sh*t sandwich,' which is if they recommend no flagships," says one senior scientist. "Most of us think that would be disastrous. The 'sh*t sandwich with a side of pickle' is when they choose no flagships but recommend technology development for whatever could come next–which is close to what happened with Astro2010. The 'nice lunch' is what we get if they pick a true flagship. And the 'perfect meal' is their picking a flagship *and* setting priorities for technology development to enable a few more."

For now, most members of the STDTs are lining up for a "perfect meal." Remarkably, after years spent extolling the virtues of their chosen missions, they have almost universally concluded that the ideal future for U.S. astronomy in space is one in which none of their pet projects triumph over all but rather where multiple flagships are somehow built and launched in rapid succession. Such an approach would effectively create a "New Great Observatories" program for the 21st century, much like the one that produced Hubble and its epochal kin. "I've promised–and many of my colleagues have, too–that if the Decadal chooses *any* flagship, then that is what I'll be cheering for," O'Meara says. "Something like the New Great Observatories can only happen if we stop shooting into each other's backyards."

Or rather the New Great Observatories can only happen if astronomers become more savvy at what Gaudi has termed "astropolitics." "I'm utterly convinced a 'New Great Observatories' program with Lynx, Origins, and LUVOIR or HabEx–a 'LuvEx,' so to speak–could be done with a single phone call to the right person," Tremblay says. "Because on Capitol Hill, it's not about total cost–it's about annual appropriation. A couple hundred million dollars a year added to NASA's astrophysics line would suffice."

Such hopeful speculations are not necessarily just wishful thinking. "We're talking a 1 or 2 percent increase in real dollars to NASA's budget to enable another Great Observatories program," says one Beltway insider. "These are the perturbations concerted advocacy can create. Only about 30 senators are really involved in appropriations, and the annual discretionary budget of the federal government is running at about $2 trillion. So divide $2 trillion by 30 and then factor in the staffers working for each of those senators. You'll find, perhaps to your horror, that anything much below about half a billion dollars a year is essentially left to staffers and lost in the margins." Tremblay puts it more bluntly. "NASA does not really work for the Executive Office of the President," he says. "It works for the 25-year-olds a few years out of college who serve on appropriations committees. A flagship mission–or a whole new series of Great Observatories–could be green-lit over lunch by some low-level staffer while they're eating a burrito."

A Single Question

To secure a multiflagship future, astronomers will likely need an overarching goal that resonates not only with professional scientists but also with policy makers and the general public they serve (burrito-munching congressional staffers included). And for that, arguably no topic has broader appeal than humankind's long, unrequited search for alien life.

"If the Decadal wants to get the most public buy-in, they should distill the next 30 years of U.S. astronomy down to a single question: How does the universe enable biology?" O'Meara says. "Think about what's needed to get answers for that. It's not just going out and taking pictures of small, temperate exoplanets. You need to track the creation of atoms and molecules from the big bang all the way to planetary biosignatures, and you need to understand how galaxies and stars arise so that the universe could make planets in the first place. You also need to understand where life can and cannot exist, which means deeper exploration of the solar system and probably even sending astronauts to Mars. Addressing that single, fundamental question could speak to all of NASA while bringing in the NSF and international partners as well."

"This is another reason to consider Astro2020 'special,'" Tumlinson says. "For the first time, we can say with a straight face that we're conceptually designing missions with reasonable expectations of finding life on planets outside the solar system. And to not do that when you have the opportunity to—well, that seems a little crazy to me."

Yet many astronomers are also uneasy about promising more than they might be able to deliver. "This mentality, that we all must be behind one big thing and that this one thing has to be so big to justify us all being behind it, is *exactly* how we've gotten into trouble before," says Sara Seager, an astrophysicist at the Massachusetts Institute of Technology, who, alongside Gaudi, co-chaired the HabEx STDT. "Of course looking for other Earths has public support! But if you put a dollar figure on it—if you tell the public it may cost a half-billion or a billion per planet, what do you think they'll say then? It just so happens that we live on a really hard planet to find—Earths are small and faint, and they're right next to big, bright stars. That makes seeing them—or searching for signs of life on them—nearly impossible."

At least, that is the case if astronomers try to achieve those goals within existing budgets and a Decadal's 10-year time frame. "For us to decide on the future of astronomy based on some arbitrary length of time that just happens to be the number of fingers we have on our hands may not be the best way to go about things," Gaudi says. "Maybe we should decide on a different, longer timespan that actually corresponds to the missions we're considering—missions that are getting more ambitious, complicated and technologically challenging over time. These really aren't 'Decadal' surveys anymore, and they haven't been for a while. They're multidecadal surveys, and we just need to start being honest about that."

After experiencing the process and its imperfections from within the National Academies, the White House and now the largest advocacy group for U.S. astronomers, Parriott offers a simple recommendation of his own. "We need to support the Decadal, because it's probably the best we can hope for in planning and executing a program of benefit to all of us," he says. "You know that old saying about democracy? Well, maybe it applies to the Decadal, too: it's the worst way to do things—except for all the other ways."

About the Author

Lee Billings is a science journalist specializing in astronomy, physics, planetary science, and spaceflight, and is a senior editor at Scientific American. *He is the author of a critically acclaimed book,* Five Billion Years of Solitude: the Search for Life Among the Stars, *which in 2014 won a Science Communication Award from the American Institute of Physics. In addition to his work for* Scientific American, *Billings's writing has appeared in the* New York Times, *the* Wall Street Journal, *the* Boston Globe, Wired, New Scientist, Popular Science, *and many other publications. A dynamic public speaker, Billings has given invited talks for NASA's Jet Propulsion Laboratory and Google, and has served as M.C. for events held by* National Geographic, *the Breakthrough Prize Foundation, Pioneer Works, and various other organizations. Billings joined* Scientific American *in 2014, and previously worked as a staff editor at* SEED *magazine. He holds a B.A. in journalism from the University of Minnesota.*

Satellite Constellations Could Harm the Environment, New Watchdog Report Says

By Jonathan O'Callaghan

Do people have a right to an unobstructed view of the heavens? For most of human history, such a question would have been considered nonsensical–but with the recent rise of satellite mega constellations, it's now being asked again and again. Mega constellations are vast groups of spacecraft, numbering in the thousands, that could spark a multitrillion-dollar orbital industry and transform global connectivity and commerce. But the rise of mega constellations also threatens to clutter the night sky, disrupt the work of some astronomers and create space debris that harms people on Earth and in space alike.

In January 2020 *Scientific American* was the first to report on a paper arguing that such constellations may be effectively unlawful because of environmental legislation enacted more than a half-century ago by the U.S. Congress. Subsequently Congress commissioned a report from the U.S. Government Accountability Office (GAO) to weigh the evidence for such claims. Released earlier this month, the report suggests that regulatory action on mega constellations is increasingly likely–and shows the high-stakes international debate over satellites' impacts on the night sky's sanctity has only just begun. Although confined to the U.S., these decisions will have far-reaching ramifications around the world, setting not just precedent for other nations but dictating whether companies can operate in the U.S. if their satellites will harm the night sky.

"Our society needs space," says Didier Queloz, an astronomer and Nobel laureate at the University of Cambridge. "I have no problem with space being used for commercial purposes. I just have a problem that it's out of control. When we started to see this increase in satellites, I was shocked that there are no regulations.

So I was extremely pleased to hear that there has been an awareness that it cannot continue like that."

The mega constellation era began in May 2019, when Elon Musk's firm SpaceX launched the first 60 satellites in its Starlink constellation. Starlink is a venture by the company to beam high-speed broadband Internet to all corners of the globe by building and maintaining a network of more than 12,000 communications satellites in low-Earth orbit. By the end of 2019, SpaceX had already launched 180 Starlink satellites. Today the constellation's numbers have swelled to more than 3,000 and account for fully half of all active satellites in space. In September 2019 Ramon Ryan, then a law student at Vanderbilt University in Tennessee, noticed how astronomers and members of the public were alarmed about the rapidly rising numbers of these satellites, all of which can be very bright in the sky when illuminated by sunlight. While bright swarms of satellites passing overhead possess a beauty all of their own, for casual sky watchers, they can be a stargaze-spoiling nuisance. And for professional astronomers, they are on the cusp of becoming an essentially unmitigated disaster, regularly photobombing the delicate observations of facilities on the ground and even ones in low-Earth orbit, such as the Hubble Space Telescope. Radio communications from these satellites can also hamper sensitive radio astronomy instruments, which require extremely "quiet" skies in order to listen in on the distant universe.

Ryan's resultant paper in the *Vanderbilt Journal of Entertainment and Technology Law* suggested that the regulatory approval of these satellites by the U.S. Federal Communications Commission (FCC) may breach environmental law as part of the U.S. National Environmental Policy Act (NEPA) enacted in 1970. Specifically, Ryan argued that the natural aesthetic of the night sky and the profession of astronomy may be protected under NEPA–but that the FCC has so far sidestepped NEPA's oversight , thanks to a "categorical exclusion" the agency was granted in 1986 (when it simply wasn't licensing that many satellites). Following coverage in *Scientific American*, a congressional staffer passed Ryan's paper to Democratic Senator Tammy Duckworth of Illinois. Duckworth then worked with her

Democratic Senate colleague Brian Schatz of Hawaii to formally request that the GAO, which audits federal agencies at Congress's behest, determine if the FCC's categorical exclusion was still valid. Furthermore, Duckworth and Schatz asked the GAO to weigh in on whether the FCC should be allowed to license so many satellites in mega constellations without NEPA review.

The results of this report were published on November 2. In the report, the GAO suggests that the FCC should revisit its categorical exclusion from NEPA and consider whether it should update its procedures in light of the rise of mega constellations. "We think they need to revisit [the categorical exclusion] because the situation is so different than it was in 1986," says Andrew Von Ah, a director at the GAO and one of the report's two lead authors. The White House Council on Environmental Quality (CEQ) recommends that agencies "revisit things like categorical exclusions once every seven years," Von Ah says. But the FCC "hasn't really done that since 1986."

According to the report's recommendations, the FCC should review whether mega constellations affect the environment, revisit its categorical exclusion from NEPA and codify what "extraordinary circumstances" would currently lead to NEPA review. The FCC has not "undertaken and documented a full examination of its categorical exclusion to ensure it is current and appropriate," the report's authors note. And while the FCC has stated it would carry out a NEPA review in the case of "extraordinary circumstances," the report notes that the agency's regulations "do not list additional factors to explain what may constitute an extraordinary circumstance." What remains unclear is whether NEPA actually applies to the "environment" of outer space–low-Earth orbit, in particular. "This is the question," Von Ah says. "We didn't opine whether it does or does not. What we were focused on was the FCC's process for making those determinations."

Von Ah says the report took more than a year to complete and incorporates representative views of industry, astronomers and the FCC itself. "It was a slightly new topic for us," he says. Karen Howard, also a director at the GAO and the other lead author, who also co-penned a companion technological assessment of mega constellations with Von Ah, says the report was "the first time we've done a technology assessment on satellite constellations and their potential environmental effects." The findings showed there were concerns in a number of areas, not just the brightness of the satellites but also the collision risk they pose in space and the possible creation of space junk, the interference to radio astronomy caused by satellite radio transmissions and even the potential for satellites reentering the atmosphere to affect Earth's climate or harm humans on the ground. "We expect a really dramatic increase in the number of satellites that has the potential to create a lot of challenges for optical astronomy and for other users of the night sky as well," Howard says. "It could potentially make it impossible to do certain studies."

The FCC declined *Scientific American*'s request for comment. In a written response to the GAO included in the latter's report, however, it stated it had reviewed that report and was "committed to ensuring that its actions, including satellite licensing activities are in compliance with the requirements of [NEPA]." It noted that the White House's CEQ was currently revising rules on NEPA regulations for federal agencies and had advised agencies to update their NEPA procedures by September 2023. "We anticipate that the FCC will conduct a review of its NEPA rules following the issuance of the revised CEQ rules, including a review of whether licensing large constellations of satellites normally does not have significant effects on the human environment," the FCC said in its response to the GAO. "We expect that, as part of that assessment, the [FCC] will consider whether to establish a timeframe and process for periodic review of our existing categorical exclusion."

The day after the GAO report's release, the FCC also announced the creation of a new bureau for its space activities, which will help the agency handle the applications for 64,000 new satellites it is presently considering. "The new space age has turned everything we know about how to deliver critical space-based services on its head," said FCC Chairwoman Jessica Rosenworcel in a statement accompanying the announcement. "The organizational structures at the agency have not kept pace as the applications and proceedings before us have multiplied, and in some cases exponentially. And you can't just keep doing things the old way and expect to lead in the new."

Bethany Johns, deputy director of public policy and interim acting director at the American Astronomical Society (AAS) in Washington, D.C., says there are a number of bills currently making their way through Congress that seek to further regulate the impact of mega constellations on the sky. "This is one step in the long march to trying to find a policy that works for everyone," she says. "It's very complex." It's all the more so, she says, because of the switch in leadership of the U.S. House of Representatives from Democratic to Republican following the recent midterm elections. The hyperpartisan legislative gridlock likely to emerge from that shift "could make finalizing policies difficult," Johns says.

For astronomy, the threat of mega constellations may be greatest for wide, deep surveys of the sky, such as those planned for the Vera C. Rubin Observatory (VRO), a $473-million ground-based telescope paired with the largest digital camera ever built that is set to switch on in Chile later this decade. The VRO's Legacy Survey of Space and Time (LSST) is designed to probe dark matter and dark energy in the universe, among many other high-priority investigations for the field. If the number of satellites in the sky rises to 50,000, in line with modest mega constellation predictions from all nations, "something like 10 percent of [VRO] images will have a satellite trail in them," says VRO chief scientist Tony Tyson of the University of California, Davis. While most science will still be possible (albeit considerably harder), light pollution from satellites may make certain investigations prohibitively difficult. Tracking fast-moving objects such as potentially hazardous near-Earth asteroids, for instance, is easily compromised by bogus alerts from gleaming satellites. "This is going to be a real showstopper for some science," Tyson says. More recent satellite applications have caused alarm, too, such as BlueWalker 3, a recently launched (and extremely bright) satellite from the Texas firm AST SpaceMobile. Designed to beam cellular broadband to the ground, BlueWalker 3 is so bright because it has a deployable antenna nearly 700 square feet in size. Future satellites from AST, known as

BlueBirds, could be twice as large and correspondingly even brighter. The FCC has yet to fully license these satellites, which, by some estimates, could outshine all other stars and satellites in the sky.

It could take months or even years for the FCC to ascertain whether its licensing of satellites warrants NEPA review, and even then it's unlikely the agency would revisit prior applications such as Starlink. The FCC's actions may be limited to simply conducting NEPA reviews for forthcoming constellations. Thousands of satellites are planned to launch in the coming years, however, not just from SpaceX but also from rival services such as Amazon's Project Kuiper and the U.K.'s OneWeb constellation (which is licensed in the U.K. but requires market access from the FCC to operate in the U.S.). So time is of the essence. The battle for the night sky continues.

About the Author

Jonathan O'Callaghan is an award-winning freelance journalist covering astronomy, astrophysics, commercial spaceflight and space exploration. Follow him on X (formerly Twitter) @Astro_Jonny.

NASA's Interplanetary Plans May Be Lurching toward Disaster

By Shannon Hall

NASA's planetary science program is in trouble. Last month NASA administrator Bill Nelson testified to Congress that the space agency's Mars Sample Return (MSR) mission urgently needs a quarter-billion-dollar infusion of extra cash–and that even more budget busting may be on the horizon. That's disturbing news, given the decades-long trend of NASA's top-priority missions ballooning in cost and wreaking havoc elsewhere in the space agency's science budget. Already MSR's "too big to fail" status is impacting other NASA projects, and rumors are flying that more severe effects still lie in store. Seeing this all-too-familiar situation unfolding yet again, some space scientists are raising the alarm in hopes of somehow shoring up vulnerable missions against the coming storm.

"We have a very ambitious planetary program," says Alan Stern, a planetary scientist at the Southwest Research Institute, who served as associate administrator of NASA's Science Mission Directorate (SMD) from 2007 to 2008. "But there's a point at which it becomes overly ambitious to the extent that it's brittle. It's not resilient to any failures."

NASA currently operates planetary science missions across the solar system–orbiting the moon, piloting rovers and a robotic helicopter on Mars, rendezvousing spacecraft with asteroids, studying the atmosphere of Jupiter and even traversing the edges of interstellar space, to cite just a few. But the two largest missions on the horizon–so-called flagships–are MSR, a daring effort to deliver Martian rocks to Earth, and Europa Clipper, a spacecraft eponymously named for the icy, ocean-harboring Jovian moon it will survey for signs of habitability.

Those missions are not going to be easy, particularly MSR, which many experts equate to multiple flagship missions in one. Already, NASA's Perseverance Mars rover is collecting samples around its landing site and dropping them in sealed, sterilized tubes for future retrieval. Now NASA and its international partners are solidifying plans for MSR's next steps. These include fetching the samples, launching them off the surface of Mars, rendezvousing in orbit around Mars with a homeward-bound transport and safely delivering the precious specimens back to Earth in a new fashion that does not risk contamination. None of these feats have been done before, and the margin for error is razor-thin. "It all has to work in a daisy chain," Stern says. "Every one of those steps will be expensive and cannot fail—or the whole daisy chain falls apart, and you don't get your samples back."

Managing and mitigating MSR's enormous technical risks is a key driver of the project's skyrocketing cost. Its allotment from NASA's annual budget has increased from $263 million to $653 million in fiscal year 2022 and then to $822 million in fiscal year 2023. Now, per Nelson's congressional testimony, it needs another $250 million to stay on track. Yet MSR has not even reached formal confirmation yet—a milestone at which NASA codifies a mission's technical details, cost and schedule baselines—meaning it is, in principle, still on the drawing board. "We've never done anything like this, so there will be surprises," says Thomas Zurbuchen, associate administrator of NASA's SMD from 2016 to 2022. "But the size of these corrections and the trends—they worry me."

NASA's finite budget is not the only fundamental resource that can be destabilized by missions that have grown too big to fail. The space agency's personnel are a factor, too, because scientists and engineers can be pulled away from smaller or more nascent projects to help support larger ones that are considered more urgent. Whether in terms of dollars or person-hours, such shuffling can amount to "robbing Peter to pay Paul," as a recent audit critiqued. When the launch date for the Perseverance rover was fast approaching, for example, NASA's Jet Propulsion Laboratory (JPL) channeled more resources into that mission—leaving MSR and Clipper depleted. Luckily, those efforts helped ensure Perseverance's successful launch, landing and subsequent surface operations on Mars. "That's because we did make those choices," Zurbuchen says. "But then we had a hard time catching up." NASA had to shift more resources to MSR and Clipper, leaving a mission to a metal-rich asteroid known as Psyche understaffed and ultimately causing it to miss its launch window. Such delays often stretch for years, incurring additional costs and further exacerbating budgetary woes all the while. "[NASA's] planetary [science division] is under financial pressure because we've snowplowed all the problems forward," Zurbuchen says.

To alleviate some of that pressure, NASA recently pushed back the launch date for a planned Venus orbiter known as VERITAS by at least three years–a decision that left many Venus scientists furious. "NASA is robbing money for a selected mission that they had baselined in their budget until as recently as last year," says Paul Byrne, a planetary scientist at Washington University in St. Louis. "It has nothing to do with anything the [VERITAS] team has done. This was essentially collateral damage."

And NASA's planetary science division isn't alone in suffering for a flagship's sins; there is usually plentiful pain to spread around. One of the most high-profile recent cases, for instance, concerned chronic overruns on the James Webb Space Telescope (JWST), a flagship mission from the space agency's astrophysics division. Originally estimated to cost $4.96 billion and launch in 2014, instead JWST launched in 2021 on a budget that had bloated to more than $10 billion. To pay for the overruns at the time, NASA terminated all participation in ExoMars, a robotic spaceflight program led by the European Space Agency and aimed at dispatching two ambitious missions to Mars (more recently, NASA rekindled some support for ExoMars). "During that time, when all that money was spent on James Webb, a lot of missions never happened," Stern says. "Or they were cancelled, or they were stillborn through no fault of their own–just because NASA had to find the resources."

Being "too big to fail," however, may be an inevitable corollary to the high ambitions scientists, policymakers and the public alike typically assign to NASA–a systemic effect perhaps best summarized by JPL's motto: "Dare mighty things." Mark Sykes, CEO and director of the Planetary Science Institute, even argues that NASA's system is *meant* to fail. Flagship missions deliberately take on tasks that defy easy execution, he says, meaning that cost estimates will always fall short. "At this time (and historically), the only buffer NASA has beyond extending these flagships out or canceling them is all the other mission and research programs, which they treat as a cookie jar to be raided as needed," he says.

It is no wonder a wide swath of researchers are concerned that MSR will similarly raid other projects outside of planetary science. In fact, it already has. NASA's 2024 budget proposal calls for delaying work on the Geospace Dynamics Constellation (a heliophysics mission) and slowing technology development for the Habitable Worlds Observatory (a future astrophysics flagship mission)—all because of MSR's hefty price tag. The agency says that it stops there—for now, anyway. "As far as within the planetary portfolio, to this point, funding for MSR has been added on top of and over and above the planetary budget," says Lori Glaze, NASA's planetary science division director. "So we have not had to take from other parts of the planetary portfolio at this point."

And yet there has been a lot of mysterious movement within many planetary science missions. VERITAS has been delayed. New Horizons, the spacecraft that gave us our first-ever detailed view of Pluto, has been moved out of the planetary science division and into the heliophysics division—effectively cutting short an "extended" phase of its mission that involved making novel astronomical measurements and potential further observations of icy bodies at the solar system's scarcely explored outer limits. NASA has also cut funds for NEO Surveyor, a space telescope designed to detect near-Earth asteroids as part of the space agency's planetary defense efforts. The agency only gave partial funding to the Dragonfly mission, an enormous nuclear-powered quadcopter meant to fly the frigid skies of Saturn's icy moon Titan. The launch window for another Venus mission known as DAVINCI was pushed back, confirms James Garvin, DAVINCI's principal investigator at the Goddard Space Flight Center. And some scientists speculate that MSR's overruns will place other Mars missions—such as the Curiosity rover—on the chopping block, too. "There's no mission under development or flying or rolling that's not at risk," Sykes says.

Glaze, however, has an explanation for all these changes. Any true delay, such as that for VERITAS, was not caused by an overrun in MSR, she says, but rather by the COVID pandemic, which introduced new inefficiencies to preparations for missions such as Europa Clipper and Psyche that had to be absorbed into the budget. In addition, NASA has been forced to reckon with rising inflation, cost growth and extremely tight supply chains. Finally, she argues that MSR is not experiencing an overrun at all because the mission has not yet reached confirmation. "There is no such thing as a cost overrun because we haven't actually set a budget yet," she says.

But Sykes disagrees, given that NASA is already spending hundreds of millions of dollars now. "Glaze's statement does not align with the facts on the ground," he says. "This is what NASA does: they play with words; they play with money so that they can allocate it; they keep things in the dark to try to keep down the criticism."

And even if MSR has not started to dip into other planetary programs, NASA itself apparently acknowledges this may only be a matter of time. In its budget proposal, the space agency stated, "Mars Sample Return costs may increase beyond the outyear profile shown in the budget, which would require either reduced funding for other science activities or descoping of this mission." For this reason, NASA is creating an independent review board that will assess the current plans—essentially providing a final check and one last chance at some kind of course correction before the agency makes a commitment. "I don't want it to impact anything else in the planetary science portfolio," Glaze says. "But remember that we do have a fixed budget. And so we have to balance everything. And that's our challenge: to try and work through how we balance all of those priorities."

"Dare mighty things," indeed. Casey Dreier, chief of space policy at the Planetary Society, notes that the extra $250 million needed this year alone will make it difficult to put VERITAS back in NASA's budget at all. And other missions, such as a time-sensitive Uranus probe that tops a representative wish list from the U.S. planetary science community, will likely never see the light of day. "The easiest mission to delay is the one that hasn't started yet," Dreier says.

To avoid such collateral damage, Stern suggests that NASA hit pause to rethink the mission's architecture or to see if foreign partners could provide resources to lighten the budgetary load. Sykes has a bigger intervention in mind: Congress. "NASA is not at all being transparent about what they're doing," he says. "Congress needs to demand that NASA–mission by mission, project by project–tell what delays have been implemented over the past year and what budget adjustments have been made." They need to shield other programs, or the fallout might just mean there is little of the planetary program left, Sykes adds. "The American taxpayer and the science of this agency is suffering as a consequence of putting everything on the altar of MSR," he says.

In response to allegations of insufficient transparency, a NASA spokesperson notes that the space agency has "posted and reported openly" on its plans for VERITAS, New Horizons and many other planetary science missions. In particular, the spokesperson notes that NASA has communicated its plans via multiple publicly available presentations to its Planetary Science Advisory Committee, community forums and "town hall" events in the past year.

This is not to say that plenty of scientists are not thrilled by the prospect of MSR, given the likelihood that its results will rewrite textbooks and could conceivably even provide the first compelling evidence of extraterrestrial life. "These samples will be studied by literally thousands and thousands of scientists for decades to come," says Laurie Leshin, JPL's current director. But at what cost? "Mars doesn't give up its secrets easily–that's been its MO for years," Dreier says. "The question will be: What price is the scientific community willing to pay–both in terms of literal dollars and other opportunities–for this mission?"

About the Author

Shannon Hall is an award-winning freelance science journalist based in the Rocky Mountains. She specializes in writing about astronomy, geology and the environment.

GLOSSARY

astrobiology A field of research on possible conditions where life can appear and evolve in places other than Earth.

biosignature A chemical or other phenomenon that may be attributed to living beings.

diminutive Of an especially small size.

dwarf star A star of smaller size and relatively lower luminosity compared to other stars; our sun is a dwarf star.

exoplanet A planet found outside of our solar system.

geyser A plume of hot water that discharges regularly from a particular source.

habitable Appropriate for life to survive in or on.

infrared A wavelength of radiation that cannot be seen as visible light, but can be used for astronomical observation.

mercurial Rapidly changing or unpredictable.

moniker An assigned name or nickname.

nucleus The center of an atom, comprising protons and neutrons, forming the bulk of an atom's mass.

organic compound A wide-ranging class of molecule that includes carbon in its makeup.

photon A particle of light or radiation that has energy but no mass.

primordial At the oldest or earliest stage, or in its most basic form.

rarefied Highly refined, elevated, or removed.

serendipitous Characterized by fortunate coincidence.

terra firma Solid ground or dry land.

ubiquitous Being widespread or easily found.

vetted Carefully investigated and proven.

wavelength The measure between high points in a wave; different wavelengths of radiation serve different types of astronomical observation.

FURTHER INFORMATION

"Infrared Astronomy," Space Telescope Science Institute, https://webbtelescope.org/webb-science/the-observatory/infrared-astronomy.

Bouzid, Samia, Carol Sutton Lewis, and the Lost Women of Science Initiative. "The Forgotten Star of Radio Astronomy," *Scientific American*, February 8, 2024, https://www.scientificamerican.com/article/the-forgotten-star-of-radio-astronomy/.

Dean, Brandi. "An Astronaut's Guide to Applying to Be An Astronaut," NASA.gov, March 2, 2020, https://www.nasa.gov/humans-in-space/astronauts/an-astronauts-guide-to-applying-to-be-an-astronaut/.

Lee, Katherine. "The Origin of the Origin of the Universe," *Astrobites*, February 4, 2023, https://astrobites.org/2023/02/04/origin-of-origin-of-universe/.

Panek, Richard. "The Most Shocking Discovery in Astrophysics is 25 Years Old," *Scientific American*, December 1, 2023, https://www.scientificamerican.com/article/the-most-shocking-discovery-in-astrophysics-is-25-years-old/.

Plait, Phil. "When We Find Earth 2.0, What's Next?" *Scientific American*, February 7, 2024, https://www.scientificamerican.com/article/when-we-find-earth-2-0-whats-next/.

Porco, Carolyn. "Cassini at Saturn: A Retrospective," *Scientific American*, October 1, 2017, https://www.scientificamerican.com/article/cassini-at-saturn-a-retrospective/.

Rees, Martin. "How Astronomers Revolutionized Our View of the Cosmos," *Scientific American*, September 1, 2020, https://www.scientificamerican.com/article/how-astronomers-revolutionized-our-view-of-the-cosmos/.

CITATIONS

1.1 Venus Might Host Life, New Discovery Suggests by Adam Mann (September 14, 2020); 1.2 Water Found in Sunlight and Shadow on the Moon by Leonard David (October 26, 2020); 1.3 Asteroid Dust from Hayabusa2 Could Solve a Mystery of Planet Creation by Jonathan O'Callaghan (December 8, 2020); 1.4 Excitement Builds for the Possibility of Life on Enceladus by Annie Sneed (June 28, 2016); 1.5 Pluto's Wonders Come into Focus by Lee Billings (March 18, 2016); 2.1 Feeling Hot, Hot, Hot: Astronomers Discover a Giant Planet Hotter Than Most Stars by Leslie Nemo (June 5, 2017); 2.2 Nearby Earth-Size Exoplanets May Have Water by Charles Q. Choi (September 2, 2017); 2.3 The Race to Find Alien Moons by Rebecca Boyle (March 1, 2021); 2.4 We Need to Widen the 'Habitable Zones' Seen around Alien Stars by Phil Plait (July 6, 2023); 3.1 The First Milky Way Black Hole Image Lets Scientists Test Physics by Seth Fletcher (May 12, 2022); 3.2 Where (Some of) Earth's Gold Came From by Imre Bartos & Szabolcs Márka (May 23, 2019); 3.3 Mystery Object Blurs Line between Neutron Stars and Black Holes by Charlie Wood (June 30, 2020); 3.4 Brown Dwarfs Could Reveal Secrets of Planet and Star Formation by Katelyn Allers (August 1, 2021); 3.5 Mysterious Fast Radio Bursts Are Finally Coming into Focus by Adam Mann (November 22, 2021); 4.1 Time Flowed Five Times Slower Shortly after the Big Bang by Lucy Tu (July 19, 2023); 4.2 Are Telescopes the Only Way to Find Dark Matter? by Chanda Prescod-Weinstein (April 1, 2022); 4.3 How Analyzing Cosmic Nothing Might Explain Everything by Michael D. Lemonick (January 1, 2024); 4.4 The Future of Gravitational Wave Astronomy by Lee Billings (February 12, 2016); 5.1 This Report Could Make or Break the Next 30 Years of U.S. Astronomy by Lee Billings (August 18, 2021); 5.2 Satellite Constellations Could Harm the Environment, New Watchdog Report Says by Jonathan O'Callaghan (November 24, 2022); 5.3 NASA's Interplanetary Plans May Be Lurching toward Disaster by Shannon Hall (May 9, 2023).

Each author biography was accurate at the time the article was originally published.

INDEX

